7

"HOW TO"
OF
SATELLITE
COMMUNICATIONS

2nd Edition

by

Dr. Joseph N. Pelton

DESIGN PUBLISHERS
800 Siesta Way • Sonoma • CA 95476 • USA

PUBLISHED BY:

DESIGN PUBLISHERS
800 Siesta Way
Sonoma, CA 95476
U.S.A.

Tel: (707) 939-9306
Fax: (707) 939-9235
Email: design@satnews•com

FRONT COVER:- ASTRA 1G
Courtesy: Societe Europeanne Des Satellites

ISBN NO.:- 0-936361-23-9

Preface

Sonoma, California
July, 1995

This 2nd edition of the "How To" book by Dr. Joseph N. Pelton, is an outstanding introduction for those of you that wish to learn more about satellite communications. I have known Joe for years, and am always amazed at the way in which he can simplify what has to be incomprehensible to the average person not involved in telecommunications.

We have come a long way since Arthur C. Clarke in 1945 dreamed up the idea of an electronic reflector circling 22,300 miles above the equator. Think of the genius that foresee such a system not only working, but in fact, ended up providing, as it does today the most efficient method for the distribution of video, voice and data.

We are now seeing the tremendous growth of Asian satellite systems with the liberalization policies of many of the Asian Telecommunications Authorities. Asia is not well cabled and consequently the competition for subscribers between cable and direct broadcast satellites has now begun in earnest. The use of new digital compression techniques has also added to the potential for both growth and competition in the satellite industry.

Reading through the twelve chapters, you will not only get a better understanding of what makes up a satellite system, but you will also get a feeling for where this technology came from and what we can expect or hope for in the future.

I wish to thank all those at Design Publishers who were involved helping Dr. Pelton getting this second edition to press. The first edition has helped thousand learn about satellites - I have no doubt that this edition will do the same.

Silvano F. Payne
Publisher

Brief Table of Contents

TABLE OF CONTENTS

CHAPTER 1

SATELLITE COMMUNICATIONS OVERVIEW

CHAPTER 2

SATELLITES: HOW THEY WORK AND HOW TO COPE WITH THEM

CHAPTER 3

THE DIGITAL SATELLITE
COMMUNICATIONS REVOLUTION

CHAPTER 4

EARTH STATION FUNDAMENTALS

CHAPTER 5

VSAT's:

CHAPTER 6

HOW AND WHY TO BUY SATELLITE NETWORKS AND SERVICES

CHAPTER 7

SATELLITES AND VOICE COMMUNICATIONS

CHAPTER 8

DATA COMMUNICATIONS AND SATELLITES

CHAPTER 9

SATELLITES AND VIDEO SERVICES

CHAPTER 10

KEY NEW TECHNOLOGIES

INTERSATELLITE LINKS AND HIGH ALTITUDE LONG ENDURANCE PLATFORMS

CHAPTER 11

LEOs AND MEOs: THE NEW SATELLITE SYSTEMS ON THE BLOCK

CHAPTER 12

THE FUTURE OF SATELLITE COMMUNICATIONS

═══ CHAPTER 1 ═══

SATELLITE COMMUNICATIONS OVERVIEW

> **"Thanks to a few tons of electronic gear twenty thousand miles above the equator, ours will be the last century of the savage, and for all mankind, the Stone Age will be over."**

Arthur C. Clarke

1.0 INTRODUCTION

Satellites are not magic. Strip away the costly and powerful launch vehicles and the little men in lab coasts who build these complex devices and you find that a communications satellite is really just a very tall microwave tower that relays telecommunications signals upon demand.

The power of communications satellite is their great height which allows coverage of vast areas.

Communications satellites can thus link together a nation, a continent, a world.

The communications satellite is a technology which is creating a global stock, bond, and money market; a host of international tele-education programs; universal television markets and ultimately the "global electronic village" (The so-called Global Information Infrastructure will not be accomplished by fiber optic cable alone).

Conventional communications satellites are positioned by rockets in space, 22,238 miles (35,786 kilometers) above the earth in order to achieve a constant "geosynchronous" orbit. (This is not reason, however, to consider satellites arcane, hard to understand or mystical). An accident of the earth's physics dictated that this special circular orbit in the equatorial plane should be at a distance almost one tenth of the way to the moon where gravitational attraction, orbital speed and a 24-hour orbit all coincide).

Today we are beginning to plan and implement new types of satellite systems which are closer to the earth's surface. These include satellites which are in so-called medium earth orbits, or MEOs (i.e. about 8000 miles or 13,000 kilometers) above the earth. We will also explore low earth orbit (LEO) satellites (i.e. 500 to 1000 miles or 800 to 1600 kilometers) over the surface of the globe. These systems produce lower transmission delays, less path loss, more frequency reuse and improved look angles for mobile satellite services in the higher latitude regions such as Europe, Canada, and Russia. In this book we will explore the ins and outs of GEO, MEO, LEO and other types of satellite systems.

In many ways it is best to forget about the "space" part of space communications". Just pretend the satellite is on a very long invisible pole that relays radio waves of very high frequencies and you begin to visualize the basic picture.

The purpose of this book is to avoid technical overkill and to still explain all the basics of communications satellites. It is in short a step-by-step road map to the world of satellites, how they work, and what services they provide. It will cover existing and future markets involving satellite applications in the areas of data, voice, video and other new areas like high definition TV. Past, present and future technology will be covered but only as useful and relevant.

Satellites represent a powerful technology that can do many things well. Terrestrial telecommunications, even fiber optic cables, cannot duplicate certain functions that satellites do superbly. Mobile services, global television broadcasting , and VSAT networks are unique satellite capabilities which fiber optics cables simply cannot equal. If you don't believe it try plugging a fiber optic cable into a jet plane or a truck, or try using cables to link together a regional network with 1000 computer nodes in it. Cables would be a hopelessly expensive way to make all the 499,500 pathways that could be created within a 1000 node system actively available. Satellites on the other hand are terribly good at creating large flexible, on-demand networks. But more about that later.

The purpose of this initial chapter is to let everyone get comfortable with satellite technology and its basic components. Let's begin with the "what", "where", "why", and "how" of satellite communications. The fundamental elements of satellite technology can be divided in two parts, the Space segment and the Ground segment. Together these two components constitute a complete communications system which can easily interconnect with ground based communications systems.

1.1 THE SPACE SEGMENT

The space segment is made up of an operational satellite plus usually a spare satellite for traffic restoration. This "spare" often carries traffic as well. Finally there is a system for keeping the satellite effectively in orbit called a Tracking, Telemetry, Command, and Monitoring (TTC&M) system.

The satellite which is the heart of the space segment has in fact only a few basic parts. There are the antennas that receive and transmit signals for communications purposes to earth stations or to relay TTC&M data and commands. There are solar cells and batteries to provide power. There are the electronic communications systems that filter, amplify, translate between uplink and downlink radio frequencies. Then there is a "platform" which provides a constant stable base from which communications can be provided to the earth below. Today the overwhelming percentage of satellites are maintained in geosynchronous orbit or as it is sometimes called the Clarke orbit. This is in honor of Arthur C. Clarke the man who first suggested in 1945 that an orbit 22,238 miles (35,786 kilometers) could be effectively used for space communications. Although Arthur C. Clarke is best known to us as the master of Science Fiction who created the HAL the mad computer in 2001 his true scientific contribution to radar, satellite communications and other fields like ocean thermal energy conversion are enormous.

In addition to the well established geosynchronous satellites, there are some low and medium orbit systems that can provide mobile voice communications, low cost store and forward messaging services as well as navigational services. There are even a very new type mega-LEO satellite systems such as the exciting new project called Teledesic. This system is backed up by the entrepreneurial team of Bill Gates of Microsoft and Craig McCaw of McCaw Cellular fame. More about these new developments later.

1.2 THE GROUND SEGMENT

The ground segment portion of satellite communications just keeps getting smaller, simpler, cheaper and better. This is the result of these key factors: (a) the space segment is becoming more sophisticated and powerful which means the ground segment can be less sensitive and the antennas lower in cost. This also means the antenna surface area can be reduced; (b) the application of new solid state semiconductor technology has further led to production costs going down and reliability going up. It's much like new color TV sets versus older ones; (c) sophisticated mass production techniques for smaller antennas have allowed new economies of scale to be produced; and (d) key innovations related to digital transmission, digital signal processing, and digital compression techniques all of which let you do more for less.

There is often some confusion about several terms that are applied to the ground segment namely earth station, antennas, VSATs, TVRO terminals, and Teleports. Each term has a somewhat different meaning.

The earth station is the entire kit and caboodle. The earth station includes the antenna, the multiplex gear for the uplinks and downlinks, the power generator, the offices for the on-site staff, spare parts, supply roads, etc. It is often true that a single earth station facility will contain several earth station antennas. When such a multi-antenna earth station facility serves a large urban area, ties together terrestrial networks and accesses several satellite systems it is considered a Teleport. Such Teleports are usually for both national and international service. Unfortunately, everyone tends to have their own definition of Teleport so you may wish to check on some others.

The antenna is of course the major communications facility needed to work to the satellite. It is typically a direct feed parabolic antenna that combines a reflector and the associated electronics as will be described later. All earth station antennas can send and receive communications. There actually many types of antennas that can be used for satellite communications. These start with simple low-gain antennas such as di-pole and Yagi units. The progress upward to helix type systems and then move on to parabolic and torus shaped antennas. The torus and parabolic systems can look directly to the satellite (i.e. a direct feed antenna) or they can have reflective focus device as in an indirect feed system. Finally there are new phased-array antennas which use lots of small electronic components which then add up to an overall antenna system. These components can be flat or conformed to any shape such as the side of an airplane or the top of a car. The technical aspects of these various types of antennas will be described later.

A TVRO is a Television Receive Only terminal. It is designed exclusively to receive TV signals. These are relatively cheap ($600 to $2500) dishes typically 1.5 to 2.5 meters in size. Larger TVRO antennas (5-10 meters and $50,000 to $100,000) support the redistribution of television programming to cable TV systems which receive the signals and then sends them to subscribers by cable. Very small TVROs (40 centimeters to 1 meter) are now being used for direct-to-the-home television via Direct Broadcasting Satellites. The Hubbard and Hughes Direct TV systems are now offering these small dishes manufactured by RCA and Thomson Ltd. for between $700 and $900 but these prices should drop considerably as volume of sales increase. A TVRO whether it be very small or much larger is always a satellite "terminal" since it "terminates" a satellite signal and cannot originate a transmission. There are also operational DBS systems in Japan, Europe, France, Germany, and the United Kingdom.

Unfortunately, usage of the phrase "terminal" is not consistently used to refer only to antenna receivers. There is certainly an important and useful distinction between a transmit/receive earth station antenna and a receive only terminal. An interactive antenna costs about 10 times more than a terminal of about the same size which can only "receive". Sometimes the "terminal" is rather imprecisely used to describe two-way antennas as well. This is the case with the phrase Very Small Aperture Terminal (VSAT). This often refers to "private" or customer premise antennas which are typically 1.0 to 2.4 meters in size and support interactive networks. Reference to two-way antennas as terminals should usually be avoided. A clear understanding of the basic difference between interactive antennas and receive only terminals sorts out those in the know from those that don't. The shape, size and technical characteristics of the ground segment is always changing, but more about that later.

1.3 SATELLITE COMMUNICATIONS SERVICES

The real purpose of satellite is of course to provide telecommunications service. Satellites are amazingly versatile because they provide such broadband frequencies, such broad coverage, and such easy interconnection among widely separated areas. Satellites are also highly cost effective particularly over very long distances or within networks. Satellites costs are insensitive to distance. Typical satellites for national, regional or even intercontinental coverage are provided in Table 1. This small but indicative list of key services is only a partial listing of the hundreds of services that now exist. Figures 1a and 1b, which follows Table 1 show in more detail which services are available in the narrow, medium and broadband ranges and the expanding areas of coverage with varying degrees of interactivity. Satellites tend to be most effective at the broad band services with the widest coverage. Terrestrial of course is best at local coverage.

A totally new technology that fills the gap between satellite and terrestrial technology, known as High Altitude Long Endurance (HALE) platforms which can provide geosynchronous type coverage but from an altitude of about 65,0000 to 70,000 feet (i.e. 18.5 to 20 kilometers). This technology which is based upon light weight, high altitude remotely piloted platforms can fly above commercial airspace and provide telecommunications for areas equivalent to South Korea or Taiwan. This new technology which can provide cost effective service over wide areas to very low cost antennas will also be addressed in subsequent chapters.

TABLE 1

NARROW BAND SERVICES

telex
telegram
facsimile (Group 3)
300 baud to 64 kilobit/second data services
low resolution slow scan electronic images
video text services
e-mail

MEDIUM BAND SERVICES

analog voice
digital voice
facsimile (Group 4)
voice plus data
voice mail
medium speed data services
(64 kilobits/second to 2.0 megabits/second)
limited-to-full motion videoconferencing
stereo /high fidelity audio
high resolution imaging
high quality video phone services
remote electronic printing
integrated digital service-ISDN quality transmission from
basic to primary rate interface (144 kilobits/second
to 2.0 megabits/second)

WIDE BAND SERVICES

analog television
digital television (6 megabits/second to 45 megabits/second
Advanced television services including enhanced definition TV
and high definition TV.
Remote Log- on and on-demand file transmission
Scientific interactive networking (using advanced work stations
with X-windows)
Interactive CAD/CAM services
Scientific Visualization (2D and 3D Displays)
Super computer interconnect

FIGURE 1A

Limited Coverage/One-way Service

Narrow Band
-Local paging services
-National, subnational & local data
broadcasts (e.g., news & stock reports)

Medium Band
-Conventional radio broadcasts
-Teletext broadcast
-Cabletext broadcast

Limited Coverage/Interactive Network

Narrow Band
-National, subnational & local
personal computer networks
-National, subnational & local
telegraph network
-National, subnational & local
audioconferencing
-National, subnational & local
electronic mail & facsimile network

Medium Band
-Local collection radio telephone
networks
-National high-speed data networks
(1.5 MBS/sec)
-National, subnational & local slow-
scan TV
-Videoconferencing
-National mobile satellite service

Broad Coverage/One-way Service

Narrow Band
-Global electronic paging
-Global or regional data broadcasts
(e.g., international or regional news
& financial systems)
-Global or regional radio distribution
-Global or regional electronic
printing of newspapers

Medium Band
-Global teletext systems
-Global data distribution systems
64 kilobits/sec. to 1.5 gigabits/sec.)

-Global or regional downloading of
computer software

Broad Coverage/Interactive Network

Narrow Band
-Global interactive data networks
(e.g., airline reservations &
electronic fund transfer networks)
-Global electronic mail & facsimile
-Global or regional
audioconferencing networks

Medium Band
-International & regional telephone
service
-International slow scan TV
videoconferencing
-Global satellite mobile
telecommunications services (e.g.
Inmarsat)
-Global or regional videotext
systems
-Global library with telephone &
data access

FIGURE 1B

WIDE BAND

Limited Coverage/One-way Service
-Lower power television
-SMATV
-MDS
-Conventional television
(VHF/UHF)
-Cable TV networks

-Local high definition TV (HDTV)
-National, subnational "electronic film" distribution
-National, subnational HDTV, DBS systems

Limited Coverage/Interactive Network
-National, subnational videophone services (e.g., picturephone)
-National, subnational & local videoconferencing/electronic meetings

-Interactive fiber optic wired city
-Interactive cable networks
(e.g., Hi-ovis project in Japan)

Limited Coverage/One-way Service
-International or regional satellite television distribution
-International or regional satellite distributed SMATV or CATV service
-International or regional DBS
-International or regional HDTV DBS service

Limited Coverage/Interactive Network
-International or regional videophone
-International or regional videoconferencing
-International DBS service with telephone or data link return channel
-International HDTV DBS service with telephone or data link return channel

As the number of services offered by satellites has increased, the quality, the reliability, the convenience, and cost performance have also generally improved. Only in the area of transmission delay have satellite services displayed a strategic disadvantage vis a vis terrestrial transmission technology. The three primary reasons why significant progress has been made in satellite communications services relate to digital communications techniques, low cost earth stations and high powered satellites with advanced antenna design and on-board switching and soon on-board processing.:

(a) <u>Analog-to-digital conversion</u> - Digital services, particularly with digital compression techniques, will tend to make communications services cheaper, more reliable, and high quality. By the year 2000, almost all urban based and business services will be digital in developed countries and a high percentage will be digital in developing countries as well. (The new Hughes DirecTV DBS satellite is able to deliver an amazing 75 channels of high quality TV per satellite simply due to new digital compression techniques.

(b) <u>Low cost earth stations</u> - Low cost earth stations will bring satellite services ever closer to the end user in terms of urban gateways, customer premise earth

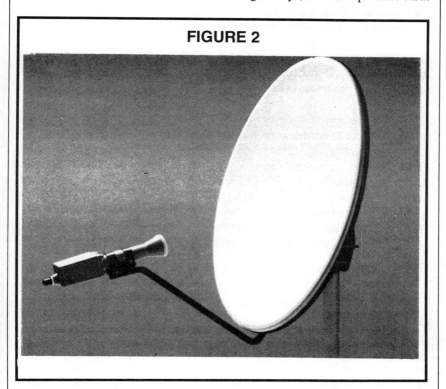

FIGURE 2

stations, desk top terminals, DBS receivers, and even hand held transceivers. In Europe, Japan, Canada, Australia, the US. and in other countries, there will be Teleports in every major city and new satellite direct broadcast antennas will be on apartment buildings, condos and in a few years millions of single family homes. Already there hundreds of thousands of interactive VSAT installed worldwide. Many multi-national companies already use desktop data reception dishes no bigger than 75 centimeters in size (see figure 2). News and financial services organizations have moved strongly into this area, like Reuters, UPI and AP. There are thousands of very small antennas built into aircraft, ships, trains, trucks and cars. There will also be smart communication beepers that can send and receive very low level but highly effective digital signals.

(c) <u>High power Ku-Band satellites</u> - Higher power and broad band satellites will be operating at the 14/12 Gigahertz (Ku-Band) and even the 30/20 Gigahertz (Ka-Band) in the 21st century as opposed to the basic (6/4 Ghz or C-Band) that gave birth to satellite communications in the 1960's and 1970's. This will provide certain advantages to sophisticated users who wish to deliver teleservices directly to urban environments but are deterred by the radio frequency environment which is very congested at the lower C-Band frequencies. Higher power and narrower antenna beams can mean higher throughput wider band service. It should also mean higher quality transmission or the use of lower cost, smaller antennas. It is important that all of these narrow and highly focussed beams be able to interconnect. This means that these advanced satellites will need on-board switching and in the next few years on-board processing to allow direct and efficient beam to beam linkages.

The bottom line is that the user community now has much greater say in the type and characteristics of the services that it can obtain via satellite. The early planners of satellite tended to be somewhat like Henry Ford in their thinking. They didn't say: "You can have any color you want just as long its black"; instead, they said if it isn't full time telephone, full time data, or occasional use television you have two options, go elsewhere or fit it into one of those three options. Earth station sizes in those days came in very limited options as well. Today flexibility for the user, service options, optimized antennas design, and high quality digital satellite services are all defining a whole different satellite environment.

1.4 HOW TO OBTAIN COMMUNICATION SATELLITE SERVICES - LEASE VS. BUY

Today, users can lease or buy satellite capacity and custom tailor it to their needs. They can buy or lease satellite capacity and with a government license they can own or lease their own earth stations. Especially in the United States almost every option is available, but things are also changing rapidly in Europe. Users can now decide a number of very key elements such as the following:

(a) Time Flexible Commitments - Contracts typically can be for a three month to two year term; for a part time/peak hour on a part time/off-peak hour occasional use commitment; or even partnership or piggy-back sharing of capacity with other users.

(b) Ownership Lease Vs. Buy - Options vary from country to country but the opportunity exists within several countries to buy satellite capacity outright for either domestic, regional or international service. In the US. you can resell parts of all of it, or broker it. One can also sub-lease it for long periods (e.g. 3-9 years), or resell it on demand by dividing the available capacity into ten or thirty minute blocks of time or more. Tax bracket, depreciation or insurance liability can be key in deciding the best strategy. Some have suggested that the US. market is "too wide open" and that the results has been to create too many unprofitable ventures, mergers and bankruptcies. What is clear today is that in the U.S. market there is a shortage of satellite capacity not only for television distribution but for the VSAT and other uses that have grown dramatically during the 1990's.

(c) Flexible Use of Capacity - In the early days of satellites you bought capacity for a specific purpose like voice or data services and that was that. Today one can buy bulk digital capacity and vary the use during the day from voice, to videoconference channel, to voice mail, to facsimile, to data. Usually voice is predominant during office hours and facsimile and data prevail at off-peak. The user now dictates how their capacity is used. As the age of Integrated Services Digital Network (ISDN) standards has arrived, the idea of having a single digital highway for all services has become increasingly common and the great flexibility of satellites to offer multiple services simultaneously has been broadly accepted. This in many ways is what the talk about the National Information Highway and the National Information Infrastructure is all about.

(d) Earth Station Antenna Characteristics - The INTELSAT system, the world's largest satellite system has well over a dozen different standards for earth stations. INTELSAT is the global communications satellite system that operates 24 satellites in the Atlantic, Pacific and Indian ocean regions. It has some 130 members and about 200 different countries and territories rely on INTELSAT for their satellite service. It also provides domestic services in some 40 countries. These antenna standard vary in size from 75 centimeters to 18 meters in cost from $2500 to $4 million. The old 30 meter stations used since the 1960s to support early, very low power satellite services are no longer being built. Domestic systems in the US., Europe and elsewhere do not have this much variation in the size and cost of their earth station antennas since the power levels and beam sizes for these more specialized satellites does not require a wide range of aperture sizes or multiplexing system.

Nevertheless there is still a wide arrange of choices available. These include: desk top antennas, TVROs, fly-away transportable, customer premise service VSATs, urban gateway antennas, large high volume earth station at international gateways, and ship to shore antennas. Today land mobile and aeronautical mobile satellite antennas have come into service. Some of these systems such as those of INMARSAT, Telesat Mobile Inc., and the American Mobile Satellite Consortium are using conventional approaches with very high powered geosynchronous satellites. Other innovators in the field, however, are planning to use unconventional low orbit satellite systems such as the Motorola Iridium cellular radio system, or the Globalstar mobile satellite system of Loral and Qualcomm. Then there are the medium earth orbit systems such as TRW's Odyssey system and the INMARSAT Project 21 (now called INMARSAT P). Other proposed systems include the Starsys system, the Ellipso system and a number of store and forward "little LEO" systems of which the Orbital Sciences Corporation's mobile satellite system known as ORBCOMM is the furthermost advanced. All of these low orbit satellites would use very small and cost effective ground transceivers. Indeed it is the improved look angle and the ability to work to very small and low cost transceivers that has largely stimulated the recent and strong interest in LEO and MEO satellites.

Collectively this explosion in the types of satellite antennas available to the user community can be summed up as the "Satellite Communications User Revolution". The rules of the game concerning the use of satellite communications have changed enormously over the last 20 years. Almost all changes have made services more affordable, more flexible, higher quality, more accessible, more controllable, and easier to own or lease for expanded periods. In short, satellites have become more user friendly. In fact, in some ways they have become user selective. Technology has helped a lot, but regulatory and institutional changes have made much of the difference. This is not suggest that all problems and challenges have been overcome. In fact, there are several key issues involving the relative performance of satellite versus fiber optics that are critical concerns. One issue is that of broad band, high quality service and the other is that of transmission latency or delay. Fiber optic cable have tremendous capacity measured in multi-gigabits/ second and extremely high quality (i.e. bit error rates of better than 10^{-10}). Satellites can, in theory, achieve high throughput rates as well and the highest capacity INTELSAT satellites can achieve a very respectable 2 to 3 gigabits/second in an all digital mode. The practical problem is that satellites need new wide band allocations to achieve very high throughput. Even more to the point, the new allocations with the necessary broadband widths are in the very difficult to deal with millimeter wave bands. The precipitation attenuation of bands of 30 GHz is extremely difficult to address so as to keep high quality service comparable to fiber. Nevertheless new antenna and processing designs such as those represented by the Hughes Spaceway design and the Teledesic satellite designs suggest that there are solutions. In short truly high powered pencil beams with on-board connectivity could be both broad band and high quality as well. Finally the low earth orbit systems and the new HALE technology give promise of systems that are faster end to end than fiber optic systems. If these two issues can be effectively overcome, it would seem that satellites can expect to make a very strong showing into the next century.

1.5 THE CONTINUING REVOLUTION IN COMMUNICATIONS SATELLITES

After Sputnik in 1957 came a rush of experimental communications satellites as shown in Figure 3. It was only in 1965, that operational communications satellites first appeared on the scene. There was the low to medium altitude Initial Defense Satellite Communications System (ISCS), the Soviet twelve hour orbit Molniya system; and the INTELSAT geosynchronous satellite system beginning with Early Bird in April1965. The only global commercial satellite service that was available was thus provided by INTELSAT. This initial service was only international and it was provided through Post and Telecommunications (PTT) organizations or designated organizations. These included the following: United States (COMSAT), United Kingdom (General Post Office now British Telecom), Italy (Telespazio), Japan (KDD), Australia (OTC), France (Directorate General Telecommunications, now France Telecom) and Canada (COTC, now Teleglobe). Commercial entities provided the service in some instances but always under monopoly arrangements. In the years that followed Canada created a monopoly commercial organization to provide domestic services. They called the new enterprise Telesat.

During the years of Presidents Kennedy and Johnson, from 1960-68, the policy leadership in the US. was focused primarily on major technological and space related innovation. In this post-Sputnik and Vietnam War era, there was considerable initiative to create a national satellite entity for domestic satellite service and to demonstrate global leadership in space systems. During the Nixon years opinion shifted toward competitive satellite systems. Despite the shift in the US. favoring competition and multiple satellite systems, most other countries proceed to implement national or regional satellite systems on the basis of traditional monopolies of the Post and Telecommunications Ministries, crown corporations or national commercial organizations with monopoly rights and privileges. The major exception to this trend has been the European ASTRA system.

FIGURE 3

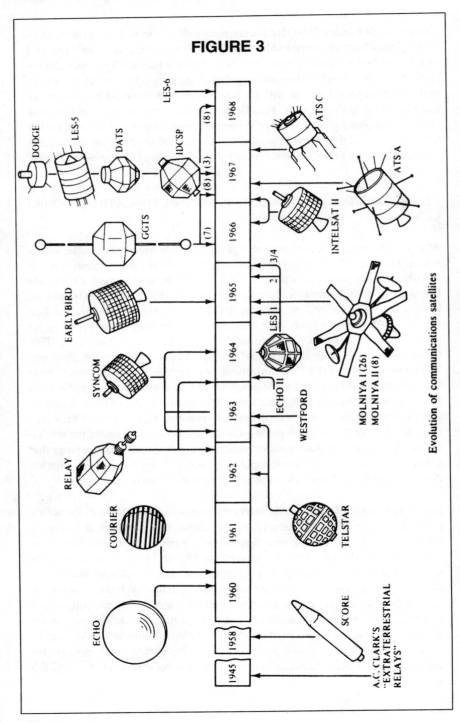

Evolution of communications satellites

The US. open skies policy was clearly at odds with the unitary approach to satellite systems of most other governments. By the mid 1980's, however, US. policy makers concluded that their open competition policy was producing benefits. They thus began to move toward deregulation, privatization and pro-competitive policies and proceeded with the divestiture of AT&T. In the UK. and Japan governments began to take a strong interest in competition and deregulation. They began to imitate US initiatives in spirit if not exactly in actual fact. Today there are major new polices on space communications in not only the US., but also Japan, Netherlands, the United Kingdom, Singapore, Australia and Germany. Perhaps the biggest backer of competition is the European Commission in Brussels, Belgium. There is one major difference between the US. and most of the rest of the world. While the US. encourages all comers, other countries have tended to limit competition to only two or three selected players. Instead of monopoly, there is "duopoly" or "triopoly"; except for the American "free for all-opoly." This too has in the United Kingdom where it's regulatory group OFTEL has decided to abandon it duopoly policy with regard to British Telecom and the Mercury Cable and Wireless and to welcome in all new, qualified competitors.

1.6 CONCLUSION

The remainder of the book will return to the themes of satellite technology, earth station technology, new services and markets, satellite cost-competitiveness versus other options, future trends, government deregulation and pro competitive policies. The constant objective will be to make the issues, the "players" and even the technical explanations as clear as a high quality satellite transmission. The first step is to understand something about satellites and how they work. This is the acid test. If you can get past Chapters 2, 3 and 4 which are somewhat of a technical challenge, the rest of the book should be as easy as an evening breeze. Alternatively if you find these chapters rough going, skip them and come back later after you have digested the less demanding parts of the book.

═══ CHAPTER 2 ═══

SATELLITES: HOW THEY WORK AND HOW TO COPE WITH THEM

> "We stand on the threshold of a new decade in which the use of satellites, digital techniques, and the introduction of VLSI technology will blur and largely eliminate the traditional boundaries between communications and information processing. This will both open up new opportunities and pose new challenges."

Dr. Joseph V. Charyk
Founding President of COMSAT

2.0 INTRODUCTION

We have already learned that satellites are radio relay transmitters in the sky, but how exactly do they work? The basic concept in Figure 4 shows us in simplified form.

Let's start by assuming the satellite is in Clarke orbit or geosynchronous orbit some 22,238 miles or 35, 786 kilometers out in space. You can round off as you like but these are the "real and precise" figures. If you want to establish your credential as an "instant expert" however these are the figures to remember. A geosynchronous satellite location means that the transmitting earth station below can almost be permanently "fixed". This means they can almost constantly use the same precise "look" angle to "see" the satellite. The receiving earth station can likewise be pointed constantly at the satellite to "hear" the satellite transmission. If the satellite is being used for two-way communications such as a telephone call or a videoconference, then the earth stations at locations A and B can both send and receive without "tracking", since the satellite is virtually motionless with respect to the earth.

In this chapter we will explore the basic "ins" and "outs" of satellite communications technology as it applies today. In chapter three which follows we will explore some of the more advanced technologies of tomorrow. This will include the technology and applications of intersatellite links (ISL's), digital compression techniques, digital signal processing, and on-board digital signal regenera-

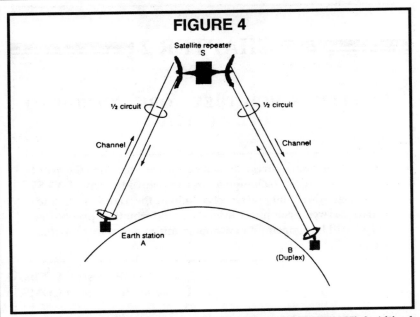

FIGURE 4

tor, low and medium earth orbit satellite systems, and the new High Altitude Low Endurance (HALE) platforms. These key technologies will serve to define the new satellites of early 21st Century. It may sound a lot like Buck Rogers' fantasy but much of the technology is coming soon and in fact much of it has flown on experimental satellites like NASA's ACTS or Japan's Experimental Test Satellites or in military satellite programs of the U.S. or Russia.

2.1 ORBITAL LOCATIONS

Earth stations have potentially great range. Using international satellites they can communicate via geosynchronous orbit to locations on the earth's surface which are up to 10,000 miles apart.

Orbital location for national satellite service is not as overwhelmingly critical as for international service. A movement of only a few degrees could totally eliminate "sight" of the international satellite when a country is at the extreme coverage limit. Mexico for instance can just "barely see" the satellite at 180E over the Pacific Ocean Region. At about 177.5 E Mexico falls off the "telecommunications cliff."

The quality of a satellite transmission is closely related to the "look angle" from the earth station to the satellite and from the satellite to corresponding earth station. This angle is determined by the earth station's longitude and latitude.

FIGURE 5

Establishing the Minimum Look Angle for an Earth Station

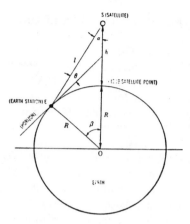

Using the above diagram as a guide, use the following formula where R equals the Radius of the earth, h equals the satellites distance above the earth's surface and the angles are those indicated.

$$\frac{R}{R+h} = \frac{\cos(\beta+\theta)}{\cos\theta}$$

The best possible look angle is 90 with a latitude which is directly below the satellite on the equator. The further away the earth station is from the satellite in terms of longitude (east or west) or from the equatorial plane (north or south latitude) the worse the look angle and the more difficult communications become. Satellite performance, the earth station, or both have to become more powerful and more expensive to achieve the same telecommunications results. Direct communications exactly to the South Pole by geosynchronous satellite for instance is impossible since a "negative look angle" results. One can add power to the satellite and install antennas with higher gain to compensate for a "look angle" problem but this does mean both more complexity and a more expensive satellite. To calculate the minimum level look angle for a satellite in orbit refer to Figure 5.

National satellite systems planners wish to obtain orbital location which put their countries as close as absolutely possible to the sub satellite point. When there are several systems seeking the same location such as when INTELSAT,

Intersputnik and the Indian domestic satellite system all were interested in the orbital location of 63 E it made for some difficulty. When such conflicting interests arise, there is a process called the intersystem coordination procedures which comes into play.

2.2 INTERSYSTEM COORDINATION PROCEDURES

Eventually, the intersystem coordination process almost always produces a reasonable compromise. Nevertheless less such solutions are sometimes difficult to devise because the stakes are so high. The difference of few degrees in the pointing angle to the satellite (or look angle) can result in cost premiums to the earth station design of lets say $3,000. If there are 2,000 stations in the network this means a $6 million premium.

When someone decides to own and operate a satellite at a particular orbital location they must obtain a formal assignment from the International Frequency Registration Board (IFRB). This is part of the overall International Telecommunications Union (ITU) which is based in Geneva, Switzerland. The ITU is a specialized agency of the United Nation Organization and actually has more members the UN. The first step in receiving an orbital assignment and frequency allocation is at the national level. If the project is a governmental satellite to be launched under the auspices of the Ministry of Post and Telecommunications or some other official governmental entity, national approval has likely already been obtained and the filing is made directly to the ITU. If it is a private venture, say in the US., the UK. or Luxembourg, then approval in the first instance must be sought from the national government. In the US. in particular, this is a tricky process since the US. "open skies" process has produced more prospective satellite operators than the old woman who lived in the shoe had kids.

This means that in the US. context there is today very definitely a lack of readily available orbital slots. This results in a process somewhat like Russian Roulette.

Everyone thus files and the FCC selects who wins and who loses. Suggestions have been made that a lottery or auction to the higher bidder would be a better process, but so far the FCC has chosen to maintain the role as Czar of orbital locations. When a private entity is proposing an international satellite service, the issues become even more complex. The US. government in this case the Federal Communications Commission (FCC) has up to this point not required private applicants for international satellite services to show they have support in other countries for the proposed service. Since the orbital locations in question can and usually will involve competing claims by other systems, this has proven to be a controversial issue that places the US. at odds with other governments. Since only one such private satellite network for international communications has been launched to date, this has been more of a theoretical rather than a practical dispute, but as the number of US. licensed international satellite organizations increases this will be an increasingly hot international issue.

Once the IFRB receives an application and clarifies any ambiguities, it is circulated to everyone in the world (i.e. all telecommunications administrations) for their comment. In particular all operators of satellite systems and those who have filed to be future operators of satellite systems can and do indicate whether there is "potential for interference".

As the number of satellites in operation has increased to over 100 and the number of satellites involved in de registration, operation or registration have increased to over 2000, the procedures to be followed had to be streamlined. Acceptance of higher levels of interference became routine, while computerization of interference calculations also became essential. The current plan for Direct Broadcast Satellites in ITU Region 2, the Americas, could not have been devised without CAOP, Computer Aided Orbital Planning.

Despite these new techniques that have helped to unsnarl the satellite orbital location logjam, the "guts" of the process I is still in the bilateral meeting between satellite systems with believe there is potential for mutual interference into each other's systems. The solutions these decisions produce often include one or more of the following:

(a) Move one or both satellites perhaps 0.5 degrees in longitude.
(b) Switch carrier assignments to move TV channels away from very sensitive channels on the other satellite.
(c) Switch off a particularly offensive transponder either permanently or until truly needed.
(d) Optimize frequency plans and transponder use schemes to achieve less interference.
(e) Take advantage of polarization differences.

These negotiated agreements that emerge from bilateral discussions generally produce the expected result: namely a rather camelesque or even "horrible compromise", but one which both can live with. There is no magic guideline to follow with regard to the intersystem coordination process but the following tips might help.

2.3 TIPS ON HOW TO SUCCESSFULLY REGISTER AND COORDINATE A SATELLITE IN GEOSYNCHRONOUS ORBIT

-Hire an Expert consultant - Try to get one of the dozen or so expert consultants that know how to get this done and have a track record (Most US. experts are in the Washington, DC area).

-File early and well - Have a good justification for your coverage areas, efficiency of spectrum use, market demand, etc. Explain why your application is unique and more meritorious than others. The FCC has now adopted the concept of a Pioneer's advantage if you are truly breaking new ground. This could be used to advantage in your filing if your applications qualifies.

-Make sure your filing is complete - Technically, operationally, financially, commercially. Also make sure your detailed schedule shows current and future expansion plan.

-Make contingency plans - If it is a US. application develop contingency plans against an FCC decision that goes contrary to your primary orbital locations or basic filing strategy. Develop good FCC contacts in order to stay well informed. Do the same with regard to the IFRB.

-Avoid black holes - If a well established satellite system has an orbital location in active use, or there is an internationally agreed plan for another use at a particular location, avoid tilting at windmills. Choose another location.

-Spend time with FCC and IFRB staff - Research with the FCC and The IFRB which locations they see as the most available and free from counter claims.

-Allow time for domestic and international processes - Filing for an FCC license, getting FCC permission to file for a location with the IFRB, and completing the IFRB coordination process can consume three years. It is for this reason, if you are indeed a US. satellite system applicant, that filing with the FCC is one of the first things to do, even if you have to go back later and amend your application.

2.4 TIPS ON BUYING A SATELLITE

Buying a satellite in some ways is like buying a paper clip; it's just that it's a thousand times more difficult and complex. The place to begin is with the service needs. As a potential buyer one must begin by asking basic questions: "What is my market?" "Do I really need a satellite to meet my market demand?" "Would a part of a satellite which I can lease or buy serve me just as well or better?" "Could I use a microwave, or fiber optic cable or some other transmission technology and meet my market demands just as well or better?" Perhaps a mix of facilities might be best?

To some, owning and operating a satellite system holds out a special mystique, that can get in the way of good business judgment. Remember, owning and operating a satellite system does not ensure a profit. More satellite systems have lost money than made money. If your plans involve selling telecommunications services your risks are likely to be higher. On the other hand, if your plan is video based your chances are probably better. If you are a US. domestic satellite applicant don't be misled. "Open skies" does not mean that US. government regulatory support will automatically follow. Consider the risks?

2.4.1 Regulatory Shifts

Take the case of the new US. private satellite systems for international service. Their market is new, exploratory, unproved and initially likely to be small. Since the cost of a new international satellite system in orbit is at least in the $150 to $300 million range there appears to be a big gap between markets, start-up costs and revenues. In short to get started the lease or purchase of transponders could serve to provide a much better chance of viability during the initial phase. Otherwise deep pockets, like in hundreds of millions of dollars of deep pockets are required.

Another problem is that several of the new start up companies believe that having their own satellite would entitle them to sell directly to users in the US., Europe and elsewhere and avoid the need for "middle men" markups. This assumption is not necessarily valid. While procedures are changing not only in the US. but also in the United Kingdom, Australia, France, the Federal Republic of Germany, and Japan. These governments are not about to allow "outside companies" to sell satellite services directly to their national citizens and businesses, but in fact there may be dozens of subtle and not so subtle roadblocks to overcome.

Nothing about owning a private international satellite systems suggests that the existing "rules of the game" are radically changed. Recall Pelton's basic Law of Telecommunications Regulation: "Most things tend to change slowly." In short when you consider going into the satellite business, consider the basic market; regulations, costs, and key options (satellites vs. terrestrial facilities; whole satellite vs. one or two transponders; buy vs. lease vs. resell vs. time-share). Don't bet on a big regulatory shift in your favor. If it does come, consider it a bonus.

There could be one gift horse in the satellite communications field which offsets the erratic nature of satellite regulatory policy. This relatively new development has considerable significance in that it offers an alternative to the purchase of a new satellite. It accordingly deserves special consideration although it will only work for some applications.

2.4.2 A New Dimension Inclined Orbit Operation—Bonanza or Bust?

The only new area of satellite technology which may still make a major and rapid impact on the field is known as "inclined orbit" operation. It is, in effect, a satellite lifetime extension technique. New approaches to satellite fuel management, to satellite-station keeping as well as a new earth station tracking techniques can change the practical lifetime of satellites from today's norm of 10 years to 15 years or longer in the future. This is sometimes a controversial subject, since the trade off to achieve the additional satellite lifetime is the need for "tracking" capability on the ground segment. This means more costly, complex and continuously controlled earth stations. This is contrary to the general trend of satellite development over the last 20 years. It also raises serious IFRB and intersystem coordination problems as well since the satellite "strays" out of the equatorial plane.

The substantial savings, in a new highly competitive environment, however, are not easily ignored. A few difficulties in regulatory procedures and a few more dollars invested in earth station tracking is small potatoes when compared to taking a $100 million satellite and extending its life from 10 to 15 years or more. Nevertheless acceptance of inclined orbit or extended operation is not universal. Skepticism remains even in the face of "megabuck" savings, even after two systems (RCA/NBC and SAFECOM) proved it works with US. domestic satellite systems. Since a fair number of the currently available transponders (i.e. satellite capacity equivalent to a TV channel) that are now offered in the US. market is "inclined orbit" capacity, a quick explanation of exactly what this means is essential.

To understand inclined orbit operation, one must begin with a better understanding of both the geosynchronous orbital arc as well as non-geosynchronous orbits as well. There is only one orbit wherein a satellite makes exactly one revolution around the earth in exactly 24 hours and maintains constant velocity and altitude by virtue of the fact that the satellite's angular momentum just cancels out the pull of gravity. Calculating this orbit is a straightforward calculus problem. It shows that at 22,238 miles (35,786 kilometers) the "local" gravitational force of .22 meters/second square is exactly canceled by a satellite traveling in stable circular orbit that is stationary to the earth below (Note the gravity here is 50 times less than on the earth's surface). But, there is a flaw to this simplistic explanation. The earth's irregular shape and density plus the gravitational pull of the moon and the sun serve to drive the satellite off of an orbit in the equatorial plan and to push it into a figure 8 shaped excursion above and below the orbital plane on a 24-hour cycle. Without intervention the excursions above and below the orbital plane tends to build up at a rate of about 0.9 degrees per year. The conventional solution has been to use hydrazine jets to keep the satellite in a tight box within 0.1 degree of the nominal location both in the east/west as well as the north/south direction. The amount of fuel required to keep the satellite on the equatorial plane, however, is ten times as great as that required for east/west station keeping.

Some satellite planners, however, have seen the potential of tracking inclined-orbit satellites. This is much easier than the sophisticated tracking procedures used with elliptical 12-hour orbits of the Soviet Union's Molniya satellite. These planners have concluded that by adding tracking to small computerized or gear driven earth station antennas they can easily maintain 24 hours a day communications with satellites even with inclinations of 5 degrees or more above or below the equatorial plane.

There are essentially three types of tracking that can be used in a satellite system. Their characteristics and relationship to extended life are given below:

(a) Continuous auto tracking

This is an expensive and sophisticated system found typically only in large master control or international gateways stations like INTELSAT Standard A or B stations. Tracking accuracy of 0.1 degrees or better is feasible with such subsystems but the cost is in the order of $35,000 to $200,000.

(b) Step tracking

This is a less precise form of tracking, which makes adjustments only in steps as the earth station "wanders" from the satellite. A combination of gears in the azimuth and axis planes serve to prevent loss of satellite acquisition. This is typically utilized on antennas 4.5 to 10 meters in size and may cost in the order of $20,000 to $30,000.

(c) Program or clock tracking

In these cases the earth station antenna doesn't really track the satellite at all. It merely points to where the satellite is "supposed" to be in accord with a detailed orbital plot against time. The build up of inclinations by a satellite without station keeping is extremely predictable since it was only applying the forces of gravity to a precisely defined physical object. In this case there is a personal computer with an appropriate program with a command capability to orient the antenna. Alternatively with clock-tracking you can simulate "tracking" with a high degree of accuracy by means of a carefully designed camshaft drive. Clock tracking is thus a similar but even more basic approach than program tracking. Mechanical cams timed to "follow" the satellite without computer guided instruction plot out a 23-hour and 56-minute celestial day. There is one problem, however. If you do not devise a system to reprogram your personal computers or reset your cams electronically, then if you ever do reposition the satellite, then you have a totally awesome job of reprogramming your earth station network especially if it is large.

There is another approach to tracking an inclined satellite that can work effectively with domestic satellite systems. This is called the "Comsat Maneuver". The solution here involves the satellite rather than the earth station. The concept is actually very simple. The satellite is put into a figure 8 like series of tilting maneuvers that during a 24-hour cycle "compensate" for the inclination buildup. If the movement above the equatorial moves the antenna beam northward, then a "tilt" brings it back down into the correct location. This works well in beams only a few degrees in size and where there is a reasonable amount of north/south margin in the coverage. Several satellites such as the SBS series in the U.S. have this technique in operation and are providing services to sophisticated users like the National Broadcasting Company (NBC). Transponders on inclined orbit satellites with up to 10 years of life expectancy on them can be purchased for under $5 million. Even more interesting to some potential consumers is the idea of putting two satellites in inclined orbit together exactly out of phase with one another to create the illusion of one unitary satellite on the orbital plane. This provides redundancy and eliminates the need for earth station tracking as long as both satellites remain in operation.

INTELSAT capacity of this type is also offered at 50% discount for the first year of operation. Such bargains should be considered, but only against market needs which may or may not be able to contend with large inclination changes. Anyway the issue of inclined orbit operation needs to be considered early since it affects the earth station antenna design, the cost of the space segment, and the types of services that can or cannot be effectively provided.

2.4.3 Off- the-shelf vs. Custom Procured Satellites

Okay, let's say you are beyond inclined orbit satellites or leasing a few high powered transponders. You know the market, your requirements and your timetable for starting services. You really need your own new satellite. The key decision to be faced at this point is whether to buy an "off-the-shelf" satellite or to develop detailed specifications for a custom designed satellite.

In the early days of commercial satellites (1965 to 1975) when the satellite manufacturing industry was just being born, the only way to get a satellite was to ask the few corporations capable of building a satellite to design to meet your needs. Since there were only a handful of satellites being built they were in effect hand crafted, one or two of-a-kind units. When INTELSAT ended up ordering eight INTELSAT III satellites from TRW in the late 1960's, it signaled the start of a new era. Such volume started corporations like Hughes and Ford Aerospace (now Space Systems/Loral) to thinking about economies of scale with basic models like airplanes if not cars. To date only Hughes Aircraft has been able to pull off the idea of mass produced satellites. Their basic spinner, the 376 series, became the Model T of the satellite industry with almost 40 of these satellites sold to Australia (AUSSAT), Brazil (SBT); Canada (TELESAT); Indonesia (PALAPA); Mexico (Morelos) plus several United States systems including AT&T(Telstar) , SBS, Western Union (Westar) and Hughes Communications itself (Galaxy). Although other manufacturers such as Space Systems/Loral,

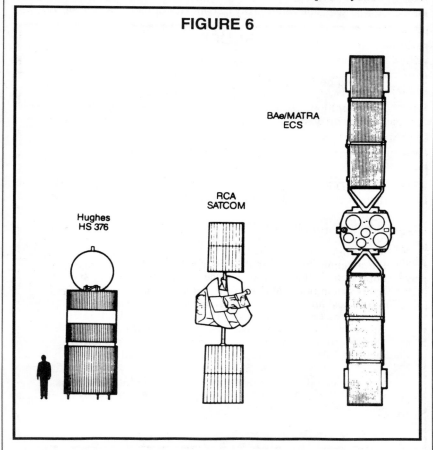

FIGURE 6

BAe/MATRA
ECS

RCA
SATCOM

Hughes
HS 376

Matra Marconi, Martin Marrietta Astro Space, TRW, British Aerospace , Aerospatiale, Mitsubishi, and Eurosatellite can also offer a basic spacecraft bus with optional antenna and communications packages, they typically have fewer basic model. See Figure 6 for representations of three types of spacecraft buses.

Even so, for purposes of most national domestic satellite communication of direct broadcast satellite programs a largely "off-the-shelf" satellite makes a lot of economic sense. Why so? In a word "money". A so called "off-the-shelf" satellite can be 30% or more cheaper. Let's compare what might be considered a typical situation of "specialized satellite" procurement versus an off-the-shelf procurement.

In this typical example the estimated cost savings are over $70 million. Furthermore since the satellite have been essentially manufactured before, most major production problems have been solved. Thus, the likely difference in the manufacturing schedule could easily be 6 months or more. In fact, an "off-the-

---

FIGURE 7

Antenna performance
Gain: The gain of the antenna is a measure of its energy concentration to or from a satellite and is given by the following equation:

$$G = \eta \left(\frac{\pi D}{\lambda} \right)^2$$

(where D is the aperture diameter of the main reflector, λ is the wavelength and η is the aperture efficiency).

Sidelobe: The unwanted spill over of radiation patterns in off-axis angles from the main beam is called a sidelobe. Maximum suppression of the sidelobes is essential to reduce the antenna noise of the earth station. This is particularly important for low elevation angle operation. In addition to the problem of the antennas' efficiency, sidelobes can be the source of interference to and from other systems, both satellite or terrestrial. The CCIR has determined the reference pattern of antenna sidelobes as 32-25 log dBi for any off-axis direction greater than one degree. A further improvement of 3 dB is recommended as a design objective.

Cross polarization: A good polarization purity of the order of 30 dB (XPD) is necessary at an earth station which accesses a satellite with dual polarization.

---

shelf" satellite might be produced in only 30 to 36 months while the custom designed satellite might require 42 to 48 months or more.

In light of these rather significant differences in terms of cost, delivery schedule and maybe even reliability, why would any one opt for a custom designed satellite? The most obvious answer of course is necessity or market advantage. Operators of say international satellite systems who are not really in a position to buy stock satellites, must clearly ask for more customized features. The complex coverage patterns for different ocean regions and the combination of international transoceanic and domestic service requirements in unique requirements for high power , limited coverage spot beams forces INTELSAT toward customization. It is also dictates medium powered global beams that can cover up to 40% of the earth at the equator are clearly a unique requirement for the global satellite system that would be useless on a domestic satellite system.

Besides international satellite systems which need spot beams and to cover up to 40% of the earth at the equator, there are also other satellite system operators who believe that certain technical innovations like operating in several different frequency bands or reusing a frequency band multiple times or having a special antenna coverage can produce market advantage. They feel that enhanced power or smaller earth station antennas can more than justify the additional cost of the satellite.

Look at it this way. When one adds in the cost of launch services, launch insurance, tracking, telemetry and command costs both during launch and deployment plus ongoing operating costs, the additional cost of the custom designed satellite shrinks to perhaps only a 10 to 15% premium. If the especially designed satellite can either double your traffic by offering unique services, or can allow you to be twice as cost efficient by say doubling your effective capacity as opposed to other satellites you get a very positive cost/benefit ratio. Figure 7 for instance shows the specialized coverage and beam interconnections on an INTELSAT 500 series. These custom-designed features do not come with off-the-shelf purchases. It is clearly important to save money but one must also remain responsive to the telecommunications market.

One word of caution, however about specialty satellites. In the mid 1980's, a satellite system manufacturer, who was also a prospective operator came up with a special design for an extraordinarily efficient satellite which achieved significantly more capacity than all competitive satellites in the US. domestic market, and did so by increasing the purchase price only modestly. The result was a satellite which was 30% more cost effective than the competition. There was, however, a serious problem in this story of satellite ingenuity. There was no demand for this huge bird in an over saturated market. The US. domestic satellite market saturated about 1985 and recovered in the early 1990's.

2.5 BUILDING A SATELLITE

But what is this thing called a satellite? Why should this gizmo that weights about a metric ton costs some $75 to $100 million, and projects an aura of Buck Rogers wizardry be worth all the fuss. In fact, the "typical satellite" today costs some $120 to $150 million or more when actually in place in geosynchronous orbit? The short answer is that despite their high costs they provide an essential service which the market wants and is willing to pay for handsomely, if need be, especially for international services. But first let's examine the high cost of satellites. There are of course may reasons why satellites are so expensive. Key among them are the following:

2.5.1 Design and Engineering

This is the biggest cost. Engineers have many challenges because the satellite must be very lightweight and built of special materials designed to reduce launch costs. It must "fold up" and squeeze into a very small volume in order to fit inside the launch vehicle. A well designed satellite must be super reliable and have redundant back-up parts because it cannot be repaired in orbit. Last but not least, it must perform a wide range of sophisticated telecommunications functions with exceptionally high quality. Million of hours of engineering and computer time can go into designing and redesigning a satellite.

2.5.2 Reliability Testing

The key to building a reliable satellite is testing, re-testing and re-testing again. Satellite components are exhaustively tested and faults examined and corrected. Then major subsystems are tested again against heat, cold, vibration, shock, lifetime wear and power surges. In space, conditions such as sunspot activity, micro meteorite bombardment, electromagnetic radiation storms, are major hazards, such events must be simulated and designs tested against them. When the subsystems have gone through tests and when everything has been redesigned and corrected the entire satellite is put through vibration shaker table, thermo-vacuum chamber, and other torture tests. Sometimes the testing part of the cycle takes over a year. Testing and redesigning consumes much of the remainder of the cost of building the satellite. Not surprising this is extremely labor intensive and also requires the use of expensive and exotic test facilities.

2.5.3 Materials and Components

The materials and components that go into building a satellite are also expensive. Often new epoxy-graphite composite materials need to be developed. Very large scale integration micro-chips, invar filters, and nickel-hydrogen batteries make for very expensive building blocks, particularly if only one in three is tested and qualified to actually be used on the satellite.

2.5.4 Other Cost Factors

The time from the start of the procurement process until the last satellite is built, launched, and has earned its full in orbit incentive payments can be a long time — 15 years or more. Satellite manufacturers must be good businessmen to avoid losing money on an overall program. They must take into account the cost of bidding the contract, amortization and modernization of the plant, test facilities, manufacturing tools, clean room, etc. They must also consider inflation and salary escalation of a large staff, non-performance of subcontractors, unexpected major design difficulties or internal or external contingencies that affect the overall program. Bidding a satellite program is extremely difficult and then managing it effectively is twice as hard. It is for these reasons that only the technologically sophisticated, large scale manufacturer with finely tuned and often near ruthless management skills can survive in this industry. The only other option is large government subsidies to aerospace companies to overcome losses. This practice which was once quite prevalent, particularly in defense related contacts, is now a decreasing practice both in the US. and abroad.

2.5.5 International Competitiveness

Today the satellite industry is as competitive as it has ever been. Several European firms like Matra Marconi, Aerospatiale, and Eurosatellite are capable of being prime contractors. Matra was a close second to Space Systems/Loral on the INTELSAT VII contract and ahead of two other US bidders. Further there are others around the world who can compete with the US. aerospace giants for satellite contracts. These include SPAR of Canada, Mitsubishi and NEC of Japan and in another decade perhaps even Chinese aerospace industry. The Indian Space Research Organization even bided on the third generation of INMARSAT satellite against four of the world's aerospace giants.

2.6 UNDERSTANDING SATELLITES

Wilbur Pritchard, a pioneer in the satellite industry once summed up the role of a communications satellite in a single sentence; "A communications satellite typically operates as a distant, but line-of-sight microwave repeater to provide communications service among multiple earth stations in widely separated geographic areas." It is not a bad definition from which to start. To understand how a satellite works you really need to understand how its various key parts are designed and operated.

2.6.1 Power Systems

Satellites must have a continuous source of electrical power 24 hours a day, 365 days a year. The number of technologies appropriate for powering satellite communications is actually quite limited. Nuclear energy is inappropriate for many reasons. These include too expensive, too much power for most mission, and too much of a liability and commercial risk. The two most logical choices which are combined on most communications satellites today are high performance batteries and solar cells.

Solar cells have much to be commended as a satellite power source. They are lightweight. They are durable and resilient to the electromagnetic radiation to which they are bombard in space. Their performance efficiency in fact only deteriorates about one to two percent per year in orbit. Their efficiency for solar energy conversion has over time edged upward from under 10% to nearly 20% with the best of the gallium arsenide cells and is expected to go much higher. Solar concentrators can also boost performance. Solar cells for satellite use are not cheap by the time they have been manufactured, tested, mounted on deployable solar panel, and totally space qualified. Nevertheless of all options among available power sources solar power is still the best, particularly for a 10 to 14 year program or longer.

There is, however, one large problem with using solar energy for a communications satellite program. The sun cannot be counted on as a continuous source of energy all year round. Twice a year a satellite in geosynchronous orbit will go into a series of 45 separate eclipses where in the sun is screened by the earth for a period which increases gradually until it is about one hour in length each day then it gradually decreases again. These solar mechanics are not to be denied and thus a supplemental on-board energy source is required. Batteries of course are the time tested answer. Initially Nickel-Cadium batteries were utilized, but more recently Nickel-Hydrogen batteries have proven to provide higher power, greater durability, and a special ability to service scores of charge and discharge cycles over the lifetime of a satellite mission. Today's typical satellite would thus likely have a one kilowatt gallium arsenide solar cell array which deploys from a three axis stabilized satellite so that it can constantly turn to achieve maximum sun exposure. It would also have sufficient Nickel-Hydrogen batteries to restore operation for the maximum predicted eclipse cycle.

Of course communications satellites are sufficiently dissimilar that there can be a number of different strategies applied to meeting power requirements. The most basic difference applies to whether the satellite is designed for point to point communications or for direct broadcast service. As a rough rule of thumb a communications satellite for telecommunications purpose will have about 1000 watts of power for each 1000 kilograms of satellite mass, while direct broadcast satellites will tend to have three times this power or 3000 watts of power for every 1000 kilograms of mass.

Since most direct-broadcast satellites are designed for limited and specific geographic coverage areas needed to provide domestic services, and since transmit programming involves a period of less than a 24-hour day there is an innovative strategy for positioning DBS satellites that can be used to meet power requirements and also "out-smart" the eclipse problem. This strategy cannot readily be applied to conventional telecommunications satellite. This strategy is to obtain a location that is sufficiently westerly so that the one hour long eclipse during the Fall and Spring equinox occurs late at night, say past midnight so that the service interruption caused by the eclipse is of minimal consequence. This strategy eliminates the need to install batteries on-board the satellite with sufficient strength to produce say three kilowatts of power at the end of the satellites life. This creates an additional savings in terms of reduced manufacturing costs for the satellite and reduced launch costs as well.

Since electrical power systems are critical to the satellite's successful operation and also consume a significant part of the mass that must be launched into orbit, there is considerable research and development underway to improve performance in this area. Rather innovative solutions for supplying power in the future now appear likely.

2.6.1.1 <u>Solar Cells</u>

Solar cells for satellite communications represents one of the more rapidly advancing areas. The first silicon solar cells of the 1960's were only about 8% efficient in converting solar energy to electricity. These gradually improved until gallium arsenide cells proved capable of achieving nearly 20% efficiency. New multiple juncture solar cells now under development should achieve close to 30% efficiency in the 1990's. Also in the 1990's we may also see increased use of solar concentrators that collect and focus the sun's rays with mirrors. These concentrators or collectors can trap and focus the sun rays so that the solar cells "see" a sun that is from three to nine times more powerful. With these new techniques, i.e. more efficient solar cells and solar concentrators it is possible to envision power systems that are several times more efficient than today's solar power systems.

2.6.1.2 <u>Batteries</u>

The Nickel-Cadium batteries of the 1960's and 1970's were relatively heavy, subject to failure, and constituted one of the prime wear out mechanisms on the satellite. After dozens and dozens of deep depth discharges these batteries had great difficulty in recovering their full charge. Today the Nickel-Hydrogen battery has been perfected and the relative performance is almost twice as great as a Ni-Cd battery because it can be discharge to 70% to 80% of its capacity. The Nickel-Hydrogen batteries are also less sensitive to the charge and discharge cycle and thus can be expected to exhibit 15 to 20 year lifetimes as opposed to 7 to 10 years of its predecessor. Finally the favorable mass to power performance of Nickel-Hydrogen cells are today almost a must. This field however is far from static. Research on new battery chemistry is progressing rapidly. New batteries based upon sulfur, sodium and even lithium all give promise of future significant gains comparable to the advantages of Nickel-Hydrogen batteries over Nickel-Cadium.

2.6.1.3 Advanced Power Concepts

Prospects for even more dramatic gains are represented by advanced fuel cell concepts, thermo-ionic converters, and ion engines that could provide propulsion, spacecraft orientation, and on-board power. Even more futuristic concepts for the late 21st century are ground based photon-beams or microwave transmissions that stabilize satellites and also provide their on-board power needs. Some research engineers feel that new advanced batteries and/or fuel cells could become so efficient that satellites could discard their solar cell wings. This does have some appeal since solar cell panels are subject to up to 90 eclipse cycles in the course of a year.

2.7 COMMUNICATIONS SYSTEMS

The communications system represents the very essence of a communications satellite. It can be divided into two key parts: the antennas and the electronic components.

2.7.1 Antennas

The antennas on the satellite have two basic missions. One is to receive and transmit telecommunications signals to support the operational mission of providing services to users. The second is to provide tracking, telemetry, command and monitoring serves in order to maintain the operation of the satellite in orbit. Of the two functions the TTC&M function must be considered the most vital. If the satellite loses communications, say because a satellite component such as a transponder fails, the situation is normally recovered in a matter of minutes or certainly hours. Usually the solution is to switch to a redundant back-up transponder or repeater. In such a case the user is inconvenienced but no irreparable harm is caused. If the vital TTC&M function is disrupted, however, then there is great danger that a spacecraft valued at anywhere from $70 million to $250 million could be permanently lost — out of control with no means of commanding the spacecraft to allow recovery. For body spinning satellites, a wobble that goes uncorrected can evolve into a "flat" spin(rather than a vertical spin). This is likely to be the ultimate disaster — total loss of spacecraft.

The antennas for communications services are the largest and most complex, while TT&C antennas are usually horn shaped or even omni-directional antennas. Let's examine the telecommunications antennas first.

2.7.2 Gain

The place to begin in understanding satellite antennas is with the concept of gain. Understand "gain" and you have gone a long way toward understanding antennas and antenna design. The most basic antenna is an omni-antenna. It transmits energy in every direction and thus disperses the signal with equal intensity in every possible receiving sector. If you have receivers located everywhere in a 360 degree radius about the antennas' X, Y and Z axes then an omni antenna makes sense. But a satellite antenna only needs to send and receive signals from earth. Furthermore, in many instances the "target" is a single high traffic catchment area say the Northeast urban corridor in the United States. If you shape an antenna to cover a few hundred square kilometers on the earth surface form a distance of 35,800 kilometers in space (i.e. the distance to geosynchronous orbit) then you must design an antenna with tremendously higher gain than an omni antenna. This produces the practical result of enhanced communications capacity and performance which is thousands of times greater. It is improved satellite antenna gain which largely explains the vast improvement in satellites today over the very first satellites. Most of the gain of an antenna in fact, represents a ratio. This ratio is a comparison of the sensitivity between the antenna being measured and an omni-directional antenna which is defined as having a base-line gain of one. Gain, because it is a ratio, is measured on a logarithmic scale expressed in decibels. The basic formula for calculating gain for parabolic antennas (i.e. the most common kind) is to calculate the square of the antenna diameter and then multiply the result by the relative efficiency of the feed and reflector system. A precise formula for calculating gain antenna sidelobe and cross polarization is provided in Figure 7.

The concept expressed by this formula is, in fact, easy to understand. The larger an antenna is, the better the lens formed to focus a tighter beam directly at the precise part of the earth to which it is intended to target the signal. A larger parabolic antenna focuses the signal beam to a much smaller area. Thus very little of the power of the transmission is diffused into space or to regions to which the satellite does not serve. The efficiency portion of the calculation then corrects for just how good the antenna's performance and design actually is. A typical satellite antenna for the Super High Frequencies such as 6/4 Gigahertz and 14/12 GHz is shown in Figure 8. Finally the higher the frequency the smaller the wavelength of the signal. This means that much more of the higher frequency signal will be trapped by a certain size of antenna than a lower frequency. Since the amount of the radio wave signal received is directly related to the area of the antenna, it means that if you either double the frequency or in effect cut the wave length in half, then the antenna gain will increase by a factor of four. Increase frequency three times and gain goes up nine times, etc.

The satellite antenna, once it receives a signal focuses it into the electronic feed system. Next filters help to eliminate unwanted signals prior to amplification. Since the received signal in geosynchronous orbit is weak (i.e. in the one thousandth to the one millionth of a watt range) amplification is essential prior to re transmission back to earth.

2.7.3 Essentials of Communications Systems

Before proceeding further it is important to learn two additional terms, namely modulation and transponder.

2.7.3.1 Modulation

Modulation is in essence a process of manipulating information in order to create a useful electronic signal. To send a signal, obviously something must be varied to supply the information content. In all forms of communications, space and terrestrial, the three aspects that are varied are typically phase (e.g. such as Phase Shift Keyed) or frequency (e.g. such as frequency modulation) or amplitude (e.g. such as amplitude modulation). The signal is usually sent on a carrier wave with the transmitting side modulating or varying the signal and the receiving side "demodulating" the signal.

There are today two basic modes of modulation schemes, namely digital modulation which varies the signal into a simple "on" or "off" code and analog modulation which creates an approximate model or "analog" of the original signal. In a way, digital coding which is geared to computer language is a direct derivative of Morse Code. Digital Modulation has many technical, economic and service advantage. These are increasing it use, particularly in developed countries. Digital modulation is highly tolerant of noise or interference since the receiver only has to determine whether the signal is on or off or, in effect, is there or not there. Digital communication is also well suited to the use of "compression techniques" which can boost satellite capacities through such techniques as Digital Speech Interpolation (DSI) and 32 and 16 kilobit per second voice service. The intricacies of how these complicated techniques operate is not

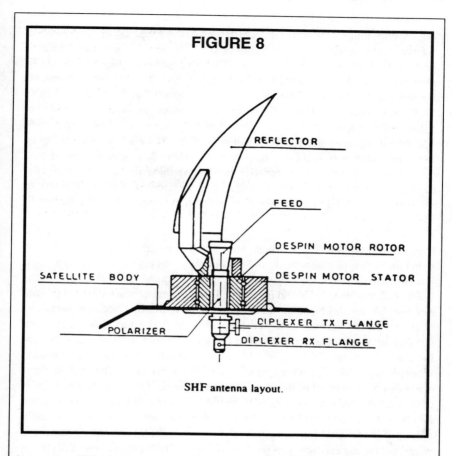

FIGURE 8

SHF antenna layout.

important. What is important is that such techniques can serve to make digital communications up to ten times or more efficient than conventional communications.

The other type of modulation is the one that has been used for over a hundred years since Bell invented the telephone. The idea is transmit via a carrier wave a signal that accurately duplicates the volume and frequency characteristics of the original signal. Amplitude modulation or frequency modulation is typically employed to send music from radio stations or telephone messages between friends. The vibration of the voice or music is directly and continuously converted into an AM or FM signal. Analog communications are thus continuously modulated or varied in response to the original signal source. It simulates the original signal source as closely as possible. Outside interference immediately affects and distorts the signal. The greater the noise in analog systems the greater the distortions and the worse the signal.

Digital communication works quite differently. It is a discrete or discontinuous form of communications. The original signal is sampled in a rapid fashion perhaps one thousand times a second. The information acquired in that milli-second or less sample is converted into a digitally encoded signal that represents the original information. The code represented by a "0" or "1" or an "on" or an "off" can be used to transmit all types of information. This can be the volume, a particular frequency or sound, light intensity or color. The digital code operates in terms of absolute information such as "0" or a "1" so there can be little confusion even in a noisy environment. There is no gradual variation in tone, volume or signal characteristics as with an analog signal. This allows the digital signal to be not only more tolerant of noise or interference, but also to facilitate the use of digital compression techniques to improve the efficiency of the transmission.

Various assumptions can be made about communications patterns and these then can be plugged into the digital communications algorithm. In scanning a line of television image, for instance, it can be "assumed" that all the dots in the line are all the same color and intensity. If it happens to be a solid background tremendous savings in data transmission can be realized. Speech patterns can also be used to streamline digital transmissions. The fact that a particular speaker in a telephone conversation talks only about 40% of the time, for instance, allows an efficiency of 2 1/2 times to be achieved. Here the technique is very straight forward. One speaker talks 40% of the time. The other speaker also talks forty percent of the time. Twenty percent of the time, the remainder of the total is filled with pauses. With digital sampling a thousand times a second it is possible to assign a line for transmission only when someone is actually talking. Since digital sampling is much faster than the human ear, no one is the wiser and the net efficiency of the system magically becomes more than twice as great.

Today there are wizards in laboratories working on new tricks of coding and digital transmission in order to do more with less. A high quality digital transmission broadcast signal for television is today considered equivalent to 68 million bits per second, even though new standards are being developed for 32 megabits/second and 45 megabits/second. Digital compression techniques based in large part upon slower sampling of the picture image has produced reasonably good videoconferencing coding and decoding equipment (CODEC) in the range 1.5 megabit per second and even lower. State of the art wizards are now even creating full motion images from a sample of "pixels" taken from those parts of the TV screen where all the action is taking place. Companies like PictureTel, Compression Labs and others are producing full color, full motion videoconference systems that operate at an amazingly low 384 to 64 kilobits per second. The 64 kilobit/second rate seems truly incredible since it is a rate a thousand times less than broadcast television and about the same as high quality voice. Sky cable, the joint venture that includes Hughes Aircraft and Rupert Murdoch's News Corporation, is currently claiming that they will be able to provide a "broadcast quality" TV channel to the home at the amazingly low digital rate of 3 megabits per second. This is largely based on new algorithms and super fast computer chips for digital processing.

This is a book about satellites and how to use them, not digital modulation and coding techniques. Nevertheless two basic messages should be communicated here. First, digital communications represents the future and it is important for anyone interested in satellite communications to also learn more about this field as well. Second, if you are a user of satellite communications it is a good thing to know a good deal about all aspects of satellites and how they work. Nevertheless it is also useful to learn how to save money on satellite communications as an overall system. Today probably the most important thing to know in the area of cost reduction is the what, why, when, where and how of choosing the right codec equipment, for your satellite network. The right codec is key to achieving significant efficiency gains in your digital transmissions. In short, efficient codec equipment may well be your best bet for bottom line cost savings.

2.7.3.2 <u>Transponders (or Repeaters)</u>

The key component in satellite communications for the operator is the device known as the transponder. The transponder is the device that takes the received signal up-linked from an earth station, filters it, translates it to the down-link frequency and amplifies it for transmission to the receiving earth station. Figures 9 and 10 show the configuration of transponders (TWTA) and the frequency plan for a 24 transponder (or 24 channel) linearly polarized GE Americom satellite.

When satellite communications began there were rather limited options available for relaying a signal across space. The first, most basic and also most inefficient was to bounce a signal off a passive reflector such as was done with the ECHO satellite. This approach could only provide extremely limited capacity. The second approach was to use a traveling wave tube (TWT) of fairly limited bandwidth, of say 36 MHz, to translate the uplink signal to the downlink signal. The trouble with these tubes which served as the "guts" of a transponder was that they were expensive, difficult to manufacture, of relative low power and subject to failure. In time, however, these tubes have gone from a few watts to 200 watts or more. Since the 1960's, however, a new option has emerged — namely the Solid State Power Amplifier (SSPA). The solid state power amplifier which utilizes the latest in transistor technology offers a number of clear advantages such as very lightweight, high reliability, long life, and flexible frequency bandwidths. There are nevertheless some difficulties with solid sate

FIGURE 9

Ge Americom communications subsystem schematic.

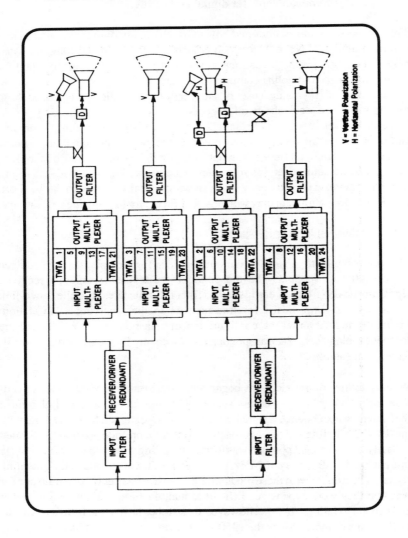

FIGURE 10

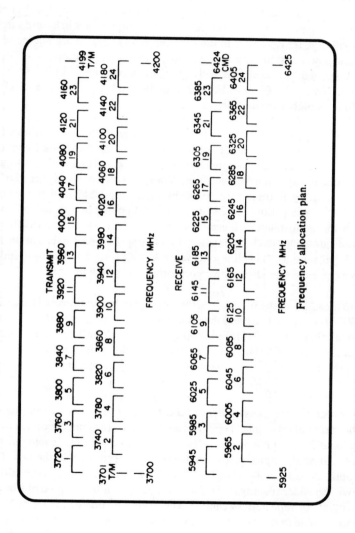

Frequency allocation plan.

power amplifiers: (a) they tend to be 5 to 10 time lower in power output than traveling wave tubes (TWTs); (b) SSPAs are about half as efficient as TWTs in performing their tasks; and (c) The relative performance of solid state amplifiers relative to TWT's drops as frequency increases.

In trading off the advantages of SSPAs versus TWTs it is not surprising that several of the latest and most sophisticated multiple frequency band satellites make use of a combination of both devices. Not surprisingly in hybrid satellites it is usually the transponders in the C-Band (6/4 GHz) that are equipped with solid state power amplifiers while the transponders for operation in the Ku-band or Ka-Band that are equipped with traveling wave tubes (TWT's). This is changing however. We will be seeing higher and higher power solid state amplifiers as higher frequencies as well.

The most common type transponder or repeater is the "quasi-linear' type wherein the uplink frequencies are separated from the downlink frequencies. This has the dual advantage of preventing oscillation and facilitating continuous and parallel transmission and reception. The translation of the higher frequency uplink into a lower frequency downlink is an important further advantage. Precipitation attenuation, the distortion of and bending of a signal as it passes through the atmosphere, especially rain storms, becomes more difficult to overcome as one moves up the radio frequency spectrum. The use of the lower frequency band as the downlink and the higher frequency band as the uplink thus serves as an excellent strategy for combating this problem.

Thus in the C-Band the uplink is 5700 to 6200 MHz and the downlink is 3700 to 4200 MHz. In the Ku-Band the uplink is again some 2000 MHz higher than the downlink, i.e. 14 GHz (uplink) and 12 GHz (downlink). In the Ka-Band the differential is 10,000 MHz apart. i.e. 30 GHz for the uplink and 20 GHz for the downlink. The expected rain attenuation at different angles is given in Figure 11.

In any event it is the transponder or repeater that truly makes a satellite work. It is the essential relay that "bend the satellite pipe" to connect point A to point B on planet Earth. Submarine cables and the new fiber optic cable must have repeaters to send a signal across an ocean. These repeaters in the new TAT-8 fiber optic cable is about 33 miles (about 55 kilometers) apart today although there will be wider spacing in the future. In fact the ultimate goal of fiber optic planners is the "repeaterless cable". The basic function of the repeater in the submarine cable or the satellite remains much the same. The repeater receives the incoming signal, filters out unwanted noise, interference and distortion and sends the amplified signal along the way toward the ultimate receiver.

FIGURE 11

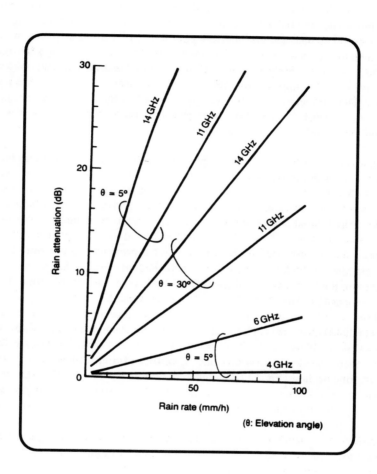

2.7.4 Stabilization

In the earliest days of satellite communications the spacecraft design was rather elegantly simple. A simple omni-directional antennas was in effect stuck on top of a coffee can like structure and spun around like a top to keep it in position. Since the antenna transmitted in all directions the earth was always illuminated even though most of the power was of course largely wasted since it was beamed off in space. Engineers quickly concluded that more sophisticated antennas would be needed to concentrate more power constantly on Earth. This meant developing a steady state platform constantly pointed toward the Earth's surface. With the INTELSAT III satellite which was first launched in 1968 this was indeed accomplished. The key was putting a spinning platform with the antennas on it inside the spacecraft platform. This antenna platform was designed to spin in exactly the opposite direction at exactly the opposite speed of about 50 revolutions per minute. The outer spinning drum exactly countered the motion of the internally spinning antenna platform. It worked marvelously well. The perfected "spinner" spacecraft could achieve antenna pointing accuracy's of 0.1 degree. Such remarkable stability and precision was far beyond what had first been anticipated.

Even so, other concepts were also explored. One of the earliest ideas was to put long boom on satellites and then use gravity gradient forces to pull the spacecraft into earth orientation. This technique proved to be relatively inefficient in terms of cost, launch constraints, deployment problems, and accuracy of orientation. Today gravity gradient techniques have been essentially abandoned.

Another technique of stabilization, however, has proven to be superior. It was the scientists of the Jet Propulsion Labs who embarked upon the first attempts to explore the rest of the Solar System. They first realized that Earth to Deep Space communications would be a highly exacting task. To send a signal from millions of miles away would require extremely exact pointing — precision of greater than 0.05 degrees. This meant a stabilization technique that was superior to the "despun" antenna. The answer way the three-axis body stabilized spacecraft. The idea was a simple one. Stabilize and orient a steady platform by putting one or more inertial or momentum wheels inside. A wheel or wheels spinning very fast could hold the satellite in a constant location just as a gyroscopic top remains in a constant position just as long as the top is spinning. A carefully engineered system with a wheel spinning at 5000 revolutions per minute has sufficient total momentum to stabilize a two metric ton spacecraft to minute tolerances.

The first three-axis stabilized satellite was the Mariner/Venus of 1962. While a long string of Mariner spacecraft continued to be built under NASA and Jet Propulsion Lab control through the 1960's it was not until the 1970's that three-axis body stabilized satellites were built for commercial communications. The spinners worked well, were reliable and cost effective. Hughes Aircraft Company, the leader in the field, believed in the "spinner' and Harold Rosen, the Hughes guru who designed Syncom and Early Bird, was not to be easily diverted to 3-axis body stabilized design. In time, however, deficiencies in the "despun" communications satellite began to emerge. These deficiencies related to design flexibility, pointing accuracy, efficiency of solar illumination of solar cell panels, and conservation of mass to reduce launch costs.

To understand these issues one only need know the differences in the basic architecture of the two types of satellites. The spinner is essentially a tin can with solar cells on the outside with a set of antenna stuck on top. The three-axis body stabilized satellite looks like a box with a group of antennas stuck on top and solar cell wings extending out from the sides of the box. Oddly enough these subtle differences have major consequences.

First of all, if one designs and builds three classes of boxes, small, medium and large, you can use these basic structures as a "bus" for a very wide range of application including meteorological and earth resource sensing spacecraft. If you need to add or subtract an antenna from the top of the bus it usually can be accommodated. The design of "spinners" tends to be much more specialized. Adding an extra antenna to the top of a spinner is not easy. It may not fit and even if it does it may easily make the entire satellite tip over and go into a flat spin. A flat spin is highly undesirable since it renders the satellite inoperable. In the area of redesign, upgrade and flexible use of a communications satellite the results are again the same. Against a 3 axis stabilized multi-purpose "bus" the spinner is a poor second.

Then, there is the issue of solar panels. The basic design of the spinner, restricts the use of solar cells to the outer shell of the satellite. Since this can be a large area, and skirts can be added to augment the power supply this is not by itself such a liability. The problem is the outside must spin around to stabilize the satellite. This means the solar cells are constantly going around and around, in and out of the sun light. On the average the solar cells see the sun about 40% of the time. In the case of three-axis body stabilized satellites an electric motor can constantly orient the solar panels in order that they are virtually always illuminated by the sun to maximum advantage. The effective output of the same number of solar cells on a spinner is thus much less than on a body stabilized spacecraft.

Thirdly, the impact of the spinner's limitations have a cumulative negative effect. More solar cells must be added to achieve the needed power but at the cost of mass and additional expense. Coping with additional antennas means not only the expense of re-engineering the spacecraft but the likelihood that the stability of the spacecraft will have to be re-established perhaps with active nutation dampers (i.e. reaction jets that can be fired to keep the satellite upright). All of these factors and more mean that the spacecraft weighs more, is more expensive to build, and then ultimately cost most a launch. It is not surprising that Hughes Aircraft the primary designer and proponent of spin stabilized satellites, has been very successful in the 1990's, with the HS 601. This is, in fact, a three-axis body stabilized satellite and is the basis of the AUSSAT satellite system. Although it is likely we will see in future years a few remaining spinners launched, the overwhelming number of future satellites for communications purposes will be body stabilized.

2.8 CONCLUSION

You should now know a good deal about satellite technology. You should also be aware of the key options to consider in terms of inclined orbit operation, spinners versus 3-axis body stabilized spacecraft and analog versus digital service for your satellite system. The pros and cons of buying "custom" or "off-the-shelf" should be particularly clear. Finally you should know the "basics" about the major subsystems of a communication satellite from solar cell to transponder to stabilization system. In short, you now probably know much about communications satellites as most Chairmen of the Board of major satellite organizations! There is still a good deal to know however about Earth station antennas, but that's another chapter.

═══ CHAPTER 3 ═══

THE DIGITAL SATELLITE COMMUNICATIONS REVOLUTION

> **"The key trends in satellite communications are digitization, intelligent transponders, upgraded and high density integration of components, and the use of artificial intelligence in software. These are also largely the keys to the C&C Society."**

Dr. Tadihero Sekimoto
Nippon Electric Corporation (NEC)

3.0 INTRODUCTION

In the previous chapter we covered the fundamentals of satellite communications as they operate today. In the fast moving field of satellite communications you can never measure progress by where you have been, but rather by where you are going. The rapid evolution of digital satellite communications is now accelerating at an increasing rate. In physics the description of accelerating, acceleration is called "jerk". Today there is no area of advanced technology in the world today that is accelerating at a faster pace. To keep up you must not only know what is happening, but what is happening next. There are more important developments now being fueled by R&D programs by NASA, DARA, CNES, ESA, ISRO, INPE, NASDA and other space agencies around the world as well as the even more important industrial telecommunications research being carried out around the world.

3.1 DIGITAL MARVELS OF SATELLITE COMMUNICATIONS

This section is hard, but it gets easier after this. At least I can assure you it is at least worth knowing about. If there is one thing to know well in today's world of satellite communications, then it is the marvels of digital communications and processing. Better, however, is not always easier and simpler to understand.

Since most applications of the future will involve digital encoding and transmission, the focus will be on the key digital technologies of the future. The key aspects of digital communications involve the conversion of an analog signal to digital format through a process called quantization. This is followed by encoding the signal and modulating it. The completion of this activity is called signal processing. This includes the multiplexing, or the combining of processed signals into an efficient carrier, and other steps needed to send a combined set of signal on a carrier form to a distant location. There is then the reversal of the process at the other end of the cycle at the receiver end. This is largely the inverse process of the activities that initiated the process. The end result is in terms of voice communications an analog-to-digital-to-analog conversion which creates what one hears as someone's voice at the end of this rather intricate process.

There are also complexities beyond these basic steps such as processing the signal within the network to "regenerate" it rather than simply receiving a signal and amplifying it along with environmental noise. Signals can be processed in very sophisticated ways. These can include combining digital processing with digital compression techniques to achieve very high rates of information rate transfer per Hz of used spectrum. With advanced digital techniques, performance as measured against this index of bits of information per Hz continue to increase steadily. We have progressed from less than half a bit of information per Hz to rates as high as three to four bits of information.

3.2 THE "INS" AND "OUTS" OF DIGITAL ENCODING TECHNIQUES

In digital encoding a signal in the form of incoming information can be simply divided into a series of "1's" and "0's" by a complete phase shift or a 180 degree rotation. This bi-phase shift keying of information can encode a signal or information with a high degree of fidelity. One can also use quadra-phase shift keying which encodes information by using only a 90 degree shift and so on. This involves a trade off between information density and throughput versus the quality of the signal. More and more information can be very efficiently encoding by using smaller phase shifts, but the chance of error and loss of signal keeps increasing. As one moves from simple bi-phase shifting to multi-phase shifting the information rate per Hz does increase and the representation of this is on an oscilloscope is seen as a series of dots of 4, 8, 16, 32, 64, or perhaps even 128. The introduction of only a very small amount of interference or noise, however, can very easily experience a total loss of signal particularly when one moves up to above 16 phase encoding.

These complex phase shifting encoding techniques are thus referred to as multi-level encoding. The addition of only a small amount of noise to the process and the resolution between the different phase shifts is immediately lost. The extremely low noise and bit error rates of fiber optic cable is considered well suited to multi-phase shift keying but the higher noise environment of free space makes this application much more challenging.

The idea of encoding of a digital signal has really two aspects. One is the mathematical concept that information is recorded as a 1 or a 0. In fact, there is actually a physical representation of that information as a pulse or non-pulse. These pulses are never really the perfect notched wave forms that are showed as theoretical models in textbooks. Such a perfectly defined depiction is only the idealized representation. The physical waves can be irregularly shaped enough to be misperceived, particularly if steps in the pulse are taken to represent a 22.5 degree, 45 degree or even a 90 degree phase shift.

The process of quantization, which starts the process of analog to digital conversion or ends it at the remote location, is actually quite simple a voice created analog wave form which looks like some variation of a singe wave. The idea is that there is a multiple step quantization scale which is used to converted rounded curve of the analog signal into a reasonably close approximation of the original signal. Instead of the signal having an infinite number of smooth steps, the new digital representation would be in error in that it would include any noise already associated with the signal and it would over represent or under-represent to a small degree most of the sampled wave form.

Nevertheless, the approximation is still very close to the original. Furthermore, once the digital wave form is created, the use of digital transmission techniques makes it largely impervious to the addition of more noise or inaccuracies in the de-quantization process in the digital to analog conversion.

Once the signal is quantized, then the encoding process is accomplished to convert the information that is the digital model of an analog voice into very simplified form. This results in a string of "1's" and "0's" which contains in compact form all the information needed to recreate the voice signal at the end of the process when the signal is converted back to voice. Once completed, then the next critical step of modulation begins.

Modulation is simply the variation of a wave form to create a signal that can be transmitted over a distance through some telecommunications media. Since the digital signal is quantized and encoded into simple "1's" and "0's" it may be the thought that this is very simple and that there might be only one or a very few ways to do this. In fact there are a very large number of analog, digital and even hybrid digital and analog techniques. There are today, however, two primary digital multiplexing techniques which are strong candidates for most digital satellite systems of the future. These are Time Division Multiple Access (TDMA) and Code Division Multiple Access (CDMA) also known more simply as Spread Spectrum. These advanced digital systems when coupled with digital compression promise expanded efficiency over today's conventional systems.

Advanced digitally based multiplexing techniques are a major way in which to increase the efficiency of communications systems. The key to measuring the efficiency of a telecommunications system is the amount of information, in the form of voice, data, or images that can be sent though it within a fixed amount of spectrum. Trying to improve this performance measure will often, however, lead to increased power consumption and increased overheads in the system. Before examining the relative merits of TDMA versus CDMA multiplexing systems, it is important to note the general advantage of digital modulation over analog systems.

There are, in fact, many advantages but three of the more important reasons are as follows: (a) The ability to use multi-phase encoding in low noise environments; (b) The ability to use intense signal processing i.e. digital compression and signal regeneration techniques either to send more information more efficiently or to combat high noise, and (c) The ability to operate successfully within a high interference and high frequency re-use environment. These factors will be discussed further below, but the third factor is particularly relevant to modulation techniques. Both TDMA and CDMA, as digitally based multiplexing systems can operate in a high co-channel interference environment. CDMA by it reliance on unique codes and spreading its signal across a broad spectrum is particularly well suited to operating in a urban environment with a high level of noise. This is true whether one is operating with ultra small aperture antennas accessing a communications satellite or within a digital cellular environment.

The idea of Time Division Multiple Access (TDMA) is quite simple. Instead of dividing up a satellite transponder or cellular frequency band by slots represented by segments of frequency, the TDMA approach is to create time slots allocated to different users. This approach allows the full power of the transmitting antenna to support a short burst of signals to be sent and then quickly transition to the next burst and then on to the next. Although we tend to think of voice, data, and video services as being continuous in nature, in a digital system this is really not so. The "1's" and "0's" are actual discrete and discontinuous events. These discrete levels are obtained by sampling a signal against a scale and then this signal is "quantized" into steps or level that approximate the analog wave form. This is followed by conversion into a digitally modulated signal based on a concept known a phase shift keying. The most common approach as noted earlier is bi-phase shift where a shift of 180 degrees (or a half turn in a circle) represents a 1. No phase shift or rotation at 0 degrees thus represents a zero.

The digital code that results from phase shift keying represents voice or data or video. It can indeed be sent in burst measured in milliseconds and processed together at the other end of the TDMA transmission. Despite the discontinuous or discrete nature of the service, it still creates the appearance of a continuous service. The separation of the carrier in the time domain is very efficient for either satellites with multiple beams to interconnect or for the various cellular systems where there are again many different beams to link together.

The idea of Code Division Multiple Access (CDMA) was first developed within the military communications systems of the United States. The motivation was, in fact, to develop a highly efficient communications that was heavily encrypted, difficult to jam, and highly robust even in a high interference environment. It turned out that CDMA was very proficient in all three regards. Its reliance on heavy processing power to extract a coded signal embedded across a broad frequency spectrum was almost hidden among the various signals against which it was overlaid. The only way to extract the wanted signal from among many other unwanted signals was to have the right code. Furthermore it was already digitally encoded so that it was in effect already encrypted.

CDMA was, however, thus not only good for military purposes but also well suited to commercial communications requirements as well. Thus, the heavy coding served to protect the signal against jamming on one hand but also to protect it against a heavy noise environment in an urban environment. Further it was also suited to work in cellular systems with more intensive levels of frequency re-use. The use of coding allows more channels to be derived by overlaying of carriers one over another. In the case of CDMA the original narrow band signal is convoluted with a unique spreading code prior to transmission. This, in effect, spreads the signal out over the entire bandwidth that is available. At the receiver end this code is then used to identify the unique message signal and "deconvolute" it. This process can thus let a narrow band signal such a voice and data be spread over a video carrier without interfering in a very perceptible wave the video signal itself.

In the case of CDMA and TDMA there is certainly higher overheads and more power consumed than in analog systems, but the net efficiency gains over analog systems clearly justify these trade-offs.. Within ten years virtually all wireless systems and satellite systems will have converted from analog based multiplexing systems such as FDMA and will have instead migrated to either TDMA or CDMA.

3.3 DIGITAL SIGNAL PROCESSING AND REGENERATION

The intersection of telecommunications and computer science took place in a real and tangible way over a decade ago. In fact, if you count the telegraph as a digital instrument, the historical relationship can be counted back over a century. The basic idea is that the key functions of telecommunications are increasingly a specific form of digital processing and that the tremendous gains being achieved in advanced processing and software development when applied to telecommunications offer major gains in performance and in cost efficiency. Furthermore the tools of telecommunications such as signaling equipment, switches, and radio transmitters are increasingly resembling computers with very specialized software.

This general trend toward universal digitalization of course has a few exceptions, but the bottom line is that telecommunications and computer processing are today among the world's fastest growing industries. These technologies are experiencing the fastest rate of productivity gain (i. e. Moore's Law), and are among the most cost efficient and profitable enterprises on the global. This parallel growth and development should be not really be surprising since the convergence process have made digital processing and telecommunications very closely inter-linked indeed.

It is in the area of signal processing that this parallelism is perhaps most clearly apparent. A digital signal processor (DSP) is, in fact, a computer that is software defined to perform specialized telecommunications functions. These functions include using generic or proprietary encoding and modulation techniques to compress and to multiplex signals for high efficiency transmission. DSP can also be used to regenerate signals at the end of the transmission-reception cycle. This process of regeneration can happen at two levels. One is to use processing power to regenerate a digital carrier wave at what might be called the macro-level. This provides clear advantage and is a function that a reasonably fast processor can do quickly and efficiently without introducing a great deal of delay or latency in the end-to-end transmission path. Typical advantage to such carrier wave processing is 1 to 3 db.

The second, much more demanding approach to signal process is what is called bit-by-bit processing. This means that the entire signal at the receive end (or on-board a satellite in the frequency translation and re-transmission process) is completely reprocessed with each bit of information being totally regenerate to create a "perfect" new source transmission. This requires a huge increase in processing power and unless a very high speed processor (i.e. a super computer) is utilized this can introduce a very high level of delay in the transmission process. If one does have the software and a super computer available the gains in very long distance transmission such as in the case of a geosynchronous satellite link can be very impressive (i.e. 6 to 9 db advantage) or 5 to 6 db for a medium earth orbit satellite.

The subject of digital compression techniques will be addressed in detail in the next section, but it is important to note here that the various forms of digital signal processing which includes multiplexing and advanced encoding, signal regeneration to create a new "source", and digital compression combine to contribute enormous efficiency gains. These collective advantages which allow new digital wireless systems to be 10 to 20 times more efficient than conventional FDMA wireless signals are indeed the most important gains in performance of all the new systems that will be addressed in this chapter. Digital signal processing is thus the central technology in all advanced satellite telecommunications systems today and a number of years into the future.

3.4 DIGITAL COMPRESSION

The single biggest factor which is currently advancing wireless telecommunications within the broad technologies that represent DSP is digital compression. It was thus seen useful to have at least a sub-section devoted to this subject.

First of all there are two basic approaches to digital compression which are typically carried out by the processors and coders / decoders (codec) located within the multiplexer. These two activities are digital speech interpolation (DSI), which obviously relates to voice services, and the other is inter frame or intra frame digital compression which can be applied to all services. Digital compression reduces the information rate required to send a service through a telecommunications system by applying complex algorithms that somehow reduce the amount of information required to complete the information exchange transaction. This could be video or imaging wherein a high degree of compression is easiest to achieve or it could also be voice, or even data where compression is most difficult to achieve.

The algorithms involving video compression are often based upon assumptions of stability of image over much of the display screen with updated information concentrating on that part of the image which changes frame to frame. In applications such as video-conferences or newscasts this technique can work very well, but in cases such as a car race or a dance contest with a great deal of motion and image changes this is much less successful. Other algorithms are based upon numbering in a code book many different complex patterns of pixels with a processor matching patterns of pixels to the closest code book equivalent.

Over the last decade the algorithms which allow digitally compressed telecommunications to become more effective have constantly improved. These techniques have allowed broadcast images to be created at levels as low a 6 megabits / second in contrast to former rates of 68 megabits per second. These systems are now fully operational in the Hughes DirecTV satellite system for instance. Videoconferencing has dropped from 3 to 6 megabits per second down to rates in the 64 kilobits to 384 kilobits per second range. Different algorithms have allowed voice transmissions to be achieved at slower and slower speeds (i.e. from 64 kilobits / second down to 4.8 kilobits / second). Even in the area of data transmissions, improvements of 20% to 30 % have been achieved.

The key to advances in this field are today based upon the use of artificial intelligence for predictive patterns to enhance compression gains. Improved performance is also being achieved through the creation of very sophisticated, processor intensive algorithms that produce very good analogies to the originals with much, less information. In the past the cost of such processors in codec equipment would have been much too expensive and introduce much too much delay to have been commercially feasible. Thus the speed of processors continues to increase while the cost falls. This means that the feasibility of even better performance in the area of digital compression increases even as codec costs seem to fall rapidly.

3.5 OTHER DIGITALLY RELATED INNOVATIONS

Advanced modulation techniques coupled with error control systems can serve to send information more securely, with high quality standards (i.e. low bit error rates), and with higher throughput rates. The leading modulation systems for satellite systems are today CDMA and TDMA—each with special advantages and liabilities. It is of interest to note that certain new modulation techniques which are under development as an optimized system for fiber optic cable such as so-called "color multiplexing", or more formally Dense Wave Division Multiplexing, is really not compatible with either CDMA or TDMA. This could eventually mean expensive and delay inducing re-modulation systems to interconnect fiber, wireless or satellite systems.

Another key area of technological development is that of advanced signaling systems. If a modern digital telecommunications system were to be considered to be analogous to a transportation system then the highways and railroad tracks would represent the broad band transmission facilities equivalent to fiber optic cables, satellites and microwave relays. The local streets would represent copper wire and cellular communications transmission towers. The switching network would be represented by clover leaf interchanges and traffic signals at intersections. The signaling system, which is the intelligence in the network, would be the traffic cops, the transportation planners, and the navigators within the cars, train, and planes. They control the routing of calls and much, more. The artificially intelligent signaling system today can provide automatic number identification, verification of line accessibility prior to connection of the bearer or voice channels themselves, call forwarding, multiple ringing patterns, etc.

The power of this "smart" signaling or network management system is significant. By more efficient routing of calls and by placing calls on the system only when they can be connected greatly expands the capacity of the network. The ability to re-route calls flexibly in case of system failures or the ability to connect roaming customers directly all serve to reduce hardware investments by substituting more cost effective software instead. Simply put, the low cost of processing power and "smart" network software can make existing telecommunications networks of all types, wire and wireless, much more efficient and reliable.

3.6 CONCLUSION

In general, digital processing techniques are an extremely powerful tools which can help satellites, ground systems, and codec and multiplexing equipment perform with higher quality, security, throughput and cost efficiency. The move toward faster, cheaper, and more compact processors (i.e. the "Cray in a shoebox") in particular should produce prodigious advances in the field of satellite communications over the next five years. The special power of the new MPEG-2 digital compression standard is today particularly prominent. The routine derivation of 18 television channels on a typical satellite transponder is spreading rapidly around the world. By 1996 there will be literally thousands of TV channels available by satellite as a result of the widespread use of compression techniques.

Digital signal processing, digital compression and digital multiplexing are certainly also helping to fuel the new surge in satellite communications with efficiency increases from 10 to even 30 times.

CHAPTER 4

EARTH STATION FUNDAMENTALS

"Today, satellite communications are indispensable for mankind as a basic tool for social activities. In this the earth segment is every bit as important as the space segment."

Dr. Kenichi Miya
Former Executive V.P. - K.D.D. Japan

4.0 INTRODUCTION

In satellite communications it takes at least three to tango. Satellite telecommunication requires at least two earth station antennas connected by a satellite in the middle. This is the simplest and in fact the least typical application of satellite technology. In this simple form, the satellite serves as little more than a bend in the cable-in-the-sky. The fascinating aspect of satellites—the element which gives them a strategic advantage over terrestrial cable systems—is their multi-destination capability. With multi-destination capabilities satellites can be linked to ever smaller and lower cost earth stations. The truth is the basic philosophy of satellite communications is the more, the merrier. It is likely that the more earth station antennas you use in a satellite network the more "value" you will derive from the system. Frequently, the benefits increase exponentially as you add earth stations to an interactive network. Consider these key types of applications where the use of multiple earth stations pays off handsomely:

- A single master earth station can broadcast TV programming to 1000, 100,000 earth terminals or even more. In Japan and the US there are now literally millions of small dishes of 45 centimeters diameter receiving as many as 150 TV channels.

- A shore earth station sending routing information via a maritime satellite can easily connect to 150 ships or more within a worldwide shipping fleet.

- A news service can send instant updates and still photos to 300 or more newspapers in dozens of countries. In fact, companies such as Reuters are using such services around the world.

- An intracorporate network can be interlinked among 37-branch locations in 16 countries with complete flexibility to use telephone, telex, voice mail, low to high speed data, and videoconferencing on demand from any location to any location. With satellite service it is of course simple to upsize or downsize the network as required. Oil companies, computer manufacturers, and manufacturers have such networks fully operational today using INTELSAT Business Services or EUTELSAT SMS services.

The prime advantages of a satellite based earth station network can be found in the following areas: (a) when the transmit or receive site is mobile (Fiber optics simply cannot provide this type of service. Try plugging a cable into a plane, ship, or a truck); (b) when there are broadcast requirements to go from one or a few points to a very large number, particularly when video broadcast services are involved; (c) when there is a large interactive network with many different pathways to be served and a need to vary service requirements between low, medium and high streams of traffic, on demand.

Take the case of a very large network of 1000 earth station antennas which all need to be interconnected at least some part of the time. Also assume for the sake of convenient calculation that the average distance between all these antenna locations is 1,000 miles. The formula to calculate the number of pathways that would be required to connect all the nodes together is given as follows:

$$P \text{ (pathways)} = N / 2 (N - 1)$$

Where N stands for the Number of Nodes in the Network.

Solving for $P = 1000 / 2 (999) = 499,500$ antenna-to-antenna pathways.

That's right, nearly a half million miles of connections. If you were to try to create a terrestrial network with fiber optic cable at 1,000 miles per pathway that would be almost 500 million miles of cable! That's a lot of cable! You would probably wait 20 years to get that much fiber optic cable if you could afford it. It is for this reason that satellite networks make a lot of sense particularly as the nodes in the system multiply.

It is for similar reasons that satellite networks of the future will have higher power and more efficient antenna beams to allow them to work effectively to smaller, lower cost antennas.

4.1 BASIC CONCEPTS

With the general background there still remains a lot of basic questions about earth stations. These include: What is an earth station? What is an antenna? What is a terminal? And how do they all work?

An earth station is a complete and self contained facility capable of sending earth-to-space transmissions and receiving space-to-earth transmissions as well. An earth station facility may include several antennas which work to multiple satellites. These antennas can and often do share common roads, shelters, power supplies, and staff. An earth station complex, particularly a large one, may be referred to as a teleport. Such a teleport may include a number of earth stations for domestic, regional and international service.

4.1.1 Antenna

An antenna can be fixed or mobile. It includes an antenna reflector which is usually a parabolic dish shape, but which can also be a toroidal shape (like an outside slice of a doughnut) or with new phase array technology can even be flat. The parabolic antenna reflectors have a direct feed focused on the middle of the antenna to receive and transmit signals. The toroidal antennas have an off-set feed that works by bouncing the signal off of the reflector and into and out of the feed system. This allows the use of more than one feed system each of which can reflect off of different parts of the single toroidal reflector. This allows one such off-set reflector equipped with multiple feed to work to more than one satellite. The illustration in Figure 12, which shows parabolic dishes, cassegrain feed antennas and offset versions of these antennas, probably make the story somewhat clearer.

Finally, there is the phase array antenna which is made up of a grouping of small solid-state amplifiers with each element in the array working like a tiny antenna and with each element electronically tuned to perform as part of an overall array. This technique was first used in radio astronomy to make multiple large scale antennas and make them work as one. Then it was used in satellite antenna designs and in military communications programs.

The feed systems in all types of antennas are connected to electronics that filter, amplify, demodulate and decode the received message, and often amplifies it again. In the reverse direction the electronics modulates and codes the message, after increasing the signal from Intermediate Frequencies (IF) to Super High Frequencies (SHF) used in satellite communications. The self-contained antenna reflector, the feed system, the electronics, the power supply, the mounting, and, if applicable, tracking system can vary enormously in size, cost, and complexity.

FIGURE 12

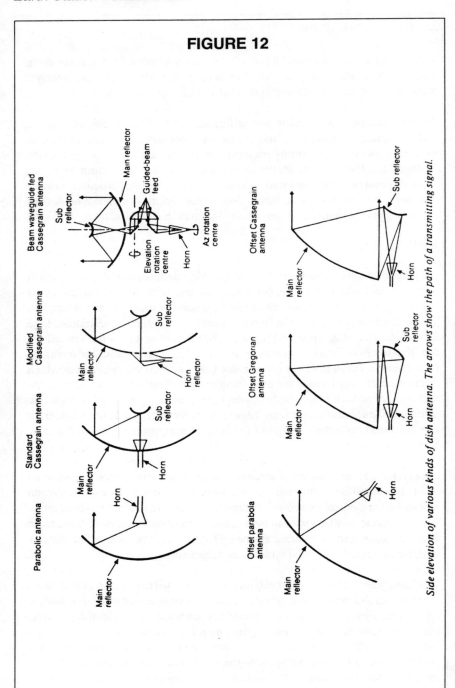

Side elevation of various kinds of dish antenna. The arrows show the path of a transmitting signal.

Within the INTELSAT system, the smallest two-way transmit and receive antenna is for low rate data (i.e., 4.8 kilobit/second). It is 1.2 meters in size and costs about $5,000. Personal communications antenna capable of providing 64 kilobit/second data or voice on a two-way basis are 1.8 meters in size and cost about $10,000. At the opposite extreme there are Standard A antennas that may be anywhere from 13 to 18 meters in size and might cost in the range of $1.2 to 3 million dollars. Clearly the larger, more complex and more expensive earth station antennas can do more, but are more appropriate for heavy streams of traffic. On the other hand the smaller, cheaper antennas may be highly appropriate for thin-route, broadcast or terrestrial bypass applications. There is a need with small antennas however, to keep quality standards as high as possible. Avoid the temptation to be penny-wise and network design foolish.

4.1.2 Terminals

This brings us to another key term frequently used for the satellite communications' ground segment-mainly the earth terminal or simply terminal. In precise usage a terminal is just that, an end point as opposed to an origination point. The most common terminals within this meaning are Television Receive Only (TVRO) terminals (3 to 7 meters in size) which now cost from $700 to $2,000 for home cable-TV type reception and $25,000 on up for cable television redistribution terminals. Receive terminals for receiving a television signal for redistribution over cable TV system have to be far more complex and sensitive than a back yard TVRO. Home mounted terminals for high-powered Direct Broadcast Satellite Systems can range from 40 centimeters to over 1 meter in size and can cost from about $250 to $900.

The other type satellite terminal is the data distribution desk-top type terminal that is about 75 cms. in size, costs about $2,000 and can only receive data albeit at the respectable rate of 9.6 or 19.2 kilobits/second. These terminals are available for both domestic and international service. The ground segment options for data distribution or even position determination is constantly expanding. Mobile satellite services and Radio Determination Satellite Services operating in the UHF (L-Band) now offers multiple options for data transmissions to extremely small microterminals that can be mounted on airplanes, ships, trains, trucks and cars. The INMARSAT standard C terminal is now under $5,000. The smallest units, such as used with the Geostar Satellite system are hand-held units that are in effect "intelligent" beepers. They weigh only a few ounces and cost under $100.

The trouble with the earth station terminology is its inconsistency. Always try to get the terms straight.

4.1.3 <u>VSAT'S</u>

Despite the technical sophistication of satellite communications, marketers have created terms that are far from clear and straightforward. Perhaps the greatest confusion was created with the introduction of the ambiguous term "VSAT". The Very Small Aperture Terminal or VSAT is a very catchy phrase meant to convey high technology, personalized satellite service, and low-cost antennas all at once. VSAT's can, in fact, be receive-only terminals but usually are two-way transmit and receive antennas capable of sending at least one voice channel plus data. Some can even transmit up to a T-1 carrier (i.e. 1.5 megabit/second in the US and 2 megabit/second in Europe).

A VSAT, to be precise, is usually a <u>VSAA</u>- a Very Small Aperture <u>Antenna</u>. When you are discussing earth station antennas and someone says "I have just the terminal for you". You should reply by saying "Are you talking about a two-way or a receive-only antenna"? This will immediately resolve what could be a basic misunderstanding. However you describe VSATs they are today a big story in the world of satellite communications. The following chapter describes VSATs and the exploding market that they represent world wide.

4.2 HOW DO EARTH STATIONS WORK?

Earth stations are earth-to-space and space-to-earth microwave relay stations. Because the distance of the radio relay in a conventional geosynchronous satellite system is 22,238 miles (or 35,786 kilometers) the propagation loss over this vast span is great. The loss is, in fact, on the order of 200 dB. Earth station must be rather sophisticated to be able to receive with high fidelity such a remote signal. The much shorter pathways of medium earth orbit and low earth orbit satellites is part of the appeal. Low earth orbit satellites can have 2500 times less path loss and 50 time less transmission delay.

Even so, everything is relative; radio telescopes must be thousands of times more sensitive than satellite earth stations to receive their signals from the stars. It's all a matter of degree.

4.2.1 <u>Radio Frequencies and Earth Stations</u>

There are many frequencies that have been allocated to satellite communications, for governmental and military uses, for aeronautical, maritime and land mobile services, for fixed commercial service, for direct broadcast services, for educational services to community antenna television, etc. The allocation chart for the use of radio frequencies for space communications is so complex that one almost needs a special engineering degree to understand which frequencies can

FIGURE 13

*Frequency allocations for satellite communications
(WARC 1979)*

Space to earth (downlink)	Earth to space (uplink)
Fixed satellite service	
2.500MHz–2.655MHz (Region 2)	
2.500MHz–2.535MHz (Region 3)	
2.655MHz–2.690MHz (Region 2. Region 3)	
3.400MHz –4.200MHz	5.725MHz–5.850MHz (Region 1)
4.500MHz –4.800MHz	5.850MHz–7.075MHz
7.250MHz –7.750MHz	7.900MHz–8.400MHz
10.7GHz –11.7GHz	12.75GHz–13.25GHz
12.5GHz –12.75GHz (Region 1 Region 3)	14.0GHz –14.8GHz
11.7GHz–12.3Ghz (Region 2)	17.3GHz –18.1Ghz
17.7GHz–21.2GHz	27.0GHz –27.5GHz (Region 2 Region 3)
	27.5GHz–31.0GHz
Maritime mobile satellite	
1.530MHz–1.544MHz	1.626.5MHz–1.645MHz
Aeronautical mobile satellite	
1.545MHz–1.559MHz	1.646.5MHz–1.660.5MHz
Mobile satellite	
1.544MHz–1.545MHz	1.645.5MHz–1.646.5MHz
19.7GHz– 21.2GHz	29.5GHz– 31.0GHz
Broadcasting satellite	
2.500MHz–2.690MHz	
11.7GHz– 12.5GHz (Region 1)	
12.1GHz– 12.7GHz (Region 2)	
11.7GHz– 12.2GHz (Region 3)	
22.5GHz– 23.0GHz (Region 2. Region 3)	

be used in each of the three regions of the world as designated by the International Telecommunications Union and the International Frequencies Registration Board (IFRB).

Figure 13 provides a listing of which frequencies are available for which services in the ITU Regions of the world. This is a bit too complicated for our purposes. In terms of basic groupings, however, it is useful to remember that the prime bands for basic Fixed Satellite Services (FSS) are the C-Band (6/4 GHz), Ku (14/12 GHz), and Ka-Bands (30/20 GHz), the prime band for Mobile Services is the VHF/UHF-bands at 1.6/1.5 GHz, and the prime bands for Military Services are the S-Bands (2.5 GHz) and X-Bands (8/7 GHz). This is of some importance to know because the antennas design and key characteristics are shaped first and foremost by the frequencies that are to be used.

It is no accident that the lower frequencies namely the L-Band frequencies were assigned to mobile communications, namely the 1.6 /1.5 GHz and the higher frequencies in the microwave band (i.e. C, Ku, and Ka-Bands) were assigned to fixed satellite services. The relatively low frequencies of the L-Band were well suited to working with simple, low-cost, light-weight, small and aerodynamically shaped antennas. This frequency is little affected by rain attenuation problems and thus there is little need for antenna performance margins that would require an antenna to perform up to 6 dB better under severe rain storm conditions. Figure 11 shows how dramatically rain attenuation effects increase with higher frequencies and low look angles to the satellite.

Several basic rules of thumb apply to the planning for earth station antenna designs that relate to the frequency to be used. At lower frequencies, somewhat ironically named the Very High Frequency (VHF) Band and the Ultra High Frequency (UHF) Band, it is appropriate to use relatively simple and inexpensive designs such as an omni-directional or Yagi or small mesh parabolic dishes. Because the wave lengths are relatively "long" the antenna does not need to be precision-tooled with very tight tolerances.

Earth station antennas operating in the microwave or Super High Frequency (SHF) range, will require much higher gain, more directionality, and precision crafted reflectors which can focus wavelengths through a high gain, precisely formed antenna. This requires more complex and expensive designs. Of course part of the antenna problem can, in part, be overcome by using much higher power levels and larger spacecraft antennas to compensate for the lack of performance on the ground.

The concept of spending more on spacecraft performance to compensate for simpler and cheaper antennas on the ground is called "Technology Inversion". In many ways this concept reflects the history of the evolution of the satellite communications industry over the last 20 years. (The satellite antennas have become larger as well as more sophisticated with higher performance. This has allowed earth station antennas to become smaller, lower in cost with increasing quality and overall performance).

4.2.2 Why Higher Frequencies?

There is an obvious question here — one which is just begging to be answered. Given the disadvantage of higher frequencies, such as higher cost and higher precision in earth station antenna curvature, problems with rain attenuation and look angles, need for higher powered and more expensive satellites, plus the need to use less proven technology, why would anyone in their right mind want to use higher frequencies?

The answer to this very logical question is three-fold. First, not everyone can farm the same land even if it is particularly fertile. The various telecommunications planners from all over the world have come together within the International Telecommunications Union (ITU) framework and agreed that they would assign the frequencies in the VHF and UHF Bands to the following types of applications—mobile satellite services of all types, land, air, and sea, radio and TV broadcasting, and terrestrial telephone service especially in developing countries. It was agreed that these types of services could best benefit from small, cheap and simple antennas. This is in part because of physical constraints (i.e., the laws of aerodynamics suggest you don't try to attach a large antenna to an airliner), or because of the economic reasons you avoid putting large antennas on cars or even ships. Since there are millions and even billions of television and radio receivers, economic efficiency suggests the broadcast antennas be as simple and cheap as possible.

This means by a process of elimination that the lower frequencies were assigned to those services that most needed smaller, less complex and cheaper antennas. This left the higher microwave frequencies in the range of 3 GHz to 30 GHz to be used for satellite telecommunications and direct broadcast satellite services, along with terrestrial microwave service as well as a number of other services. It was agreed by these international planners that these services could better adapt to the use of the higher frequencies. Nevertheless, it is also true that the lowest Fixed Satellite Service (FSS) frequency band, i.e., the 6 and 4 GHz band are preferred for use by developing countries, while the more economically advanced developed countries are tending to use the 14 and 11 GHz range more and more. Only a very few countries, namely Japan Italy, and the United States are actively using the 30 and 20 GHz band which of course, is currently the most technically challenging.

The first answer is thus that higher frequencies were assigned to satellite communications simply because there is just not enough room in the lower bands. The second reason, however, is lot more positive-namely a lack of congestion. In truth there are some attractive aspects to using the SHF frequency bands that must not be overlooked. The lower frequency bands are today largely congested and heavily shared with terrestrial telecommunications services, like HF and VHF radio telephone, VHF and UHF terrestrial television, and land based mobile communications services. This means special precautions must be taken to eliminate interference between terrestrial and space services.

Earth station antennas in the early days of satellite communications frequently were put in out of the way locations well away from the urban centers. The first locations for INTELSAT earth stations were Etam, West Virginia, Andover, Maine, and Goon Hilly Downs in the United Kingdom. Later when antennas were moved closer to the cities, elaborate screening was erected at great expense. In New York City, the Staten Island Teleport has sunk antennas into vast hollowed out pits to avoid interference .

There are also other techniques such as the use of a special modulation and coding technique known as "spread spectrum" that can help to work around this difficult problem. Clearly, flexible antenna location is a big problem at the lower frequencies even though cellular radio telephone systems at 800 MHz and soon 2 GHz has taught us that interference problems in urban environments can be overcome. Today, unwanted interference is still virtually a non-problem at 30 and 20 GHz, while even Ku-Band service still gives reasonable flexibility in the US. In Japan and Europe, however, the 14 and 12 GHz Bands are starting to be heavily filled. In another decade the surging demand for wireless applications will likely saturate the C, Ku, and Ka-Bands for fixed, mobile and DBS applications. The future will therefore force satellite systems and earth stations to move into the 30 to 50 GHz Bands.

The third and final reason in favor of higher frequencies for space communications is the potential of new antenna beams that may allow the reuse of frequencies 20, 30 even 100 times. Also new on-board information processing can allow the design of advanced "intelligent" "Ku" Band and "Ka" Band satellites of awesome capabilities. Higher frequencies satellites are better suited to the application of intelligence on board the satellite especially when the very wide band frequencies of 30 and 20 GHz are opened up for use. The Ka-Band satellites of the 21st century have enormous potential. This process is like trying to exploit a huge unused virgin territory—a new frontier where few settlers have yet ventured and where major technological break through are not only possible, but highly likely.

In the highest microwave band for space communications now allocated, namely the 30 and 20 GHz Bands, there is thus many times more spectrum available than in the lower bands (The higher the frequency the greater the available bandwidth to use). There is virtually no competing ground use and few other satellite operators are interested in using the same frequencies in nearby orbital locations. This directly translates into operational and economic benefits. First of all, you can locate an earth station antenna almost anywhere, on the top of building, in a parking lot, or even on top of a desk with a view toward the equatorial plane. Second, there is little likelihood of a problem with orbital assignments, intersystem coordination, or modification of satellite or earth station design to accommodate other satellite systems. With new high-powered Traveling Wave Tubes (TWT's) and soon high performance Solid State Power Amplifiers (SSPAs), it is or soon will be possible to build satellites that blast tremendous amounts of power down to the earth with margins large enough to withstand heavy rainstorms and relatively low-look angles. This indeed defines extremely well the design of the 30/20 GHz Teledesic mega-LEO satellite system.

With advanced antenna designs such as those used on the Advanced Communications Technology Satellite (ACTS) the permissible limits on satellite downlink power will be easily reached. These beams can also be duplicate many dozens of times over to create targeted beams for regions or even individual cities or ultimately even specific high traffic station antennas. As we see the evolution of multi-beam high frequency satellites we will also see more intelligent satellite capable of signal regeneration and information-processing on-board the satellite as well. The advanced concepts defined in experimental programs such as NASA's ACTS satellite, NASDA and MPT's ETS VI satellite and the European Space Agency's Artemis Project have defined how such satellite with on-board processing can be done. Now the Motorola Iridium Project and Teledesic are moving toward commercial implementation.

What does all this mean for the earth station antenna? It means that the size and performance of the ground antenna will reflect the inverse of the technical complexity of the satellite. High satellite power will mean smaller earth station antennas of lower gain. Intelligence on-board the satellite will mean that the terrestrial part of the system can be "dumber". It also means that power margins can be even lower.

In the area of direct broadcast satellites, roof-top 12 GHz antennas as small as 40 cms. have been tested in Japan, and the new flat antennas, developed in the US by COMSAT and manufactured in Japan Matsushita offer even better performance potential. In time these DBS antennas could be combined in function with Fixed Satellite Service (FSS) in order to perform functions like paying bills, ordering from a catalog, or getting custom newspaper printouts. In general it can be concluded that use of frequencies for satellites are going up the frequency spectrum, and that as power, multi-beam antennas, and on-board intelligence is added to the satellite, the size, cost and complexity of earth stations will go down. This favorable cost trend will be further accentuated by the use of improved solid state amplifiers, improved construction materials, enhanced manufacturing techniques, and new phase array designs.

4.2.3 Satcoms and the Frequency Problem

Satellite and earth station antenna technology are developing rapidly. This is a good thing too, otherwise the satellite frequency congestion problem could get out of hand. We are seeing positive developments that help alleviate the pressure on satellite frequencies. A surge of use of radio frequencies for US satellite communications began in the 1970's with the "Open Skies Policy" This explosion of demand served as the signal of a potential problem. Fortunately, the burgeoning demand for more frequency in the geosynchronous orbit brought about several simultaneous developments: (a) Frequency reuse by multiple antenna beam transmission; (b) Use of new and higher frequencies; (c) Polarization discrimination to allow frequency re-use; and (d) digital compression techniques. The more recent surge in applications for LEO and MEO satellites who must coordinate their frequency use with GEO systems has tended to offset these key technical gains.

(a) Reuse

If you are designing a geosynchronous satellite to serve say the entire Atlantic Region, one available option is to use a large antenna reflector and offset feeds to create two geographically separated antenna beams, one to cover North and South America, and another to cover Europe and Africa. Since the two beams are physically isolated by thousands of miles, the exact same frequencies can be used twice. This approach was, in fact, first used in the INTELSAT services in the 1970's. Now there are ways to use larger antennas and higher powered beams to up to six isolated beams and thus achieve a six-fold frequency reuse. In time, with space platforms or satellite clusters, we may see twenty-fold reuse and ultimately 100 fold reuse in Ka-Band. This technique, just like franchising McDonald hamburger stands is excellent for multiplying your coverage as your customer service demands grow. It is an approach that also complements frequency re-use through polarization discrimination (See (c) below).

This same approach to frequency reuse can be used even more effectively with MEO and LEO systems. The closer a satellite is to the earth's surface, the easier it is to create a tightly focused geographic beam. The Motorola Iridium system and the Teledesic network anticipate using 50 separate beams to achieve very high levels of frequency re-use.

(b) New Frequencies

A number of system operators in the 1970's and 1980's built new satellites that operated in the Ku-Band or some even built hybrid Ku-Band and C-Band systems that could be cross-strapped on-board the satellites. This meant that a C-Band earth antenna could up-link at 6 GHz but down-link to a Ku-Band antenna at 12 or 11 GHz. Likewise a Ku-Band antenna could up-link at 14 GHz but down-link at 4 GHz. As we look to the 1990's the Ka-Band is starting to receive increased commercial attention. The success of the ACTS test and demonstration carried out by NASA using the 30/20 GHz Band has led to a number of new Ka-Band filings such as the Norris satellite system, the Cellsat system, and Teledesic In the 1980's the Ku-Band were the prime challenge, but the Ka Band is clearly the challenge of the 1990's. The Ku-Band antennas (both in space and on the ground) were smaller in diameter (or aperture size) since the frequencies were higher and the wave length to be "shaped" was smaller.

On the other hand the accuracy of the curve on the Ku-Band reflector had to be much more precise. Thus, the use of the Ku-Band frequency in a sense was a dead heat on a merry-go-round. The C and Ku-Band antennas were, in effect, after the engineering was done about the same in cost—neither more nor less expensive to manufacture. The Ku-Band antenna performance, however, turned out to be a bigger problem than had first been anticipated. Power levels had to be turned up to cope with heavy rainfall and the distortion of the signal thundershowers caused. The late 1990's will thus likely see transition from Ku-Band to Ka-Band. This will mean that there will be attempts to achieve a very similar transition to smaller but more precisely from antennas with even higher tolerances. This means that antennas of the future in these higher bands will start to be as precise as a lens.

There is another approach, however, and this is the new style phased array antenna. In this case a number of small electronic components are placed together in an interlocked grid and then integrated to work as a total unit. This means that the antenna is "electronically formed" with a processor which creates the virtual reflector lens on demand. Ultimately phased array antennas can "electronically track" mobile units and they could be molded to any surface flat or convoluted since the processor can create any "virtual" shape that then antenna requires.

(c) Polarization Discrimination

Another big innovation to help save frequency spectrum was to introduce polarization discrimination techniques into satellite transmission. This is the "good" type of discrimination rather than a form of bias. To understand this concept you only need think of Polaroid sun glasses. The light in the case of the sunglasses is "discriminated" into two categories. "Wanted" light can flow in and out, but unwanted glare is blocked from entering. This sort of "light screen" can also work with radio frequencies. If you use two polarizers exactly out of phase with one another, you can double the use of the frequency bands by separating the transmissions to and from the satellite into "wanted" and "unwanted" receptions. You can, thus, in effect double its capacity but getting radio frequencies to work twice for you in parallel but completely out of synchronization.

There are, in effect, two approaches that can be followed. The first and most common is dual orthogonal polarization. The polarized signals are sent at right angles to one another. This technique is most common because it is the simplest and cheapest to do. Most domestic and regional satellite systems use this approach. INTELSAT and INTERSPUTNIK systems use the more unusual technique known as left and right hand circular polarization. In this case instead of "screening" on the basis of putting waveforms at right angles, the signals are circularized in the right hand direction and opposed by left hand circularized waveforms. The performance of this system is quite good, but the polarizers are more expensive.

Circular polarization is particularly effective with offset antennas whereby the signal is "reflected" indirectly into the receiving dish rather than straight in as in a "direct feed antenna". The discrimination between the "wanted" signal and the "suppressed" signal can be very high to give good communications performance. In most satellite systems the minimum level of discrimination—the ratio of wanted to the unwanted signal is often at least 24 to 25 db, while ratios can be as high as 30 db or more has been set in some satellite systems. The ITU has recommended levels for polarization discriminations that applies to both linear and circular techniques.

If one relaxes polarization purity too far, it starts to interfere and becomes a "dirty" service that pollutes other carriers. Accordingly, a relaxation of polarization is usually closely controlled and minimized to the greatest extent possible.

(d) Closer Orbital Spacing of Satellites

Another technique which has been used to expand the amount of frequency capacity available for satellite communication is simply to allow the satellites to be more closely positioned together along the equatorial orbital plane in geosynchronous orbit. Initially it was thought that satellites would need to be located some 5 degrees apart to avoid radio wave interference. This was equivalent to about 2400 miles (3805 kilometers) of physical separation. In time this distance has been reduced and in the US. the FCC, in fact, has reduced the required spacing to only two degrees or 970 miles. (1550 kilometers). This change did not come without a penalty. Smaller earth stations which tend to emit unwanted side-lobe transmissions off of the axis of the intended transmission.

This meant that such small aperture antennas had to be redesigned to limit their interference into adjacent satellite systems. Such actions, however, serve to double the number of permissible satellite systems. One approach to this problem is to distort the antenna reflectors shape so that the unwanted sidelobes tend to be distributed in the north-south axis of the equatorial plane. At least for geosynchronous satellites, the interference sensitivity is along the equatorial belt in the east-west direction. With the advent of LEO and MEO systems which travel north to south, the problem of interference returns.

(e) Digital Communications and Digital Compression Techniques

Digital compression techniques are much different from the other approaches which "expand" the frequencies that are available, i.e., frequency reuse, new frequencies and polarization discrimination, and tighter orbital positioning. Digital communications techniques, nevertheless, can also be a major factor in expanding the practical utilization of the frequency spectrum. These increases can be significant. In fact the practical gains can easily be as much as 5 to 10 times. In short, digital communications, combined with digital compression techniques are currently the biggest multiplier of "effective" capacity. Further digital signals are more robust and less subject to interference than are analog signals. The details of how these digital processing advantages are obtained were discussed in greater detail in the preceding chapter.

In short, the combination of new frequencies, closer orbital spacing of satellites, new frequency utilizations, frequency re-use by geographic isolation and polarization discrimination and digital compression techniques has effectively served to expand the practical usable capacity available through global satellite systems by more than 100 fold. If one looks to the future in terms of new frequency allocations in the millimeter wave bands, on-board processing of cellular beam patterns with hundreds of cells and advanced digital processing it is possible to foreseen even a thousand-fold increase over the next 20 to 30 years. In short the technical issues related to the availability of satellite frequencies can likely be solved for some decades to come. The increasing policy and economic problems of the "auctioning" of frequencies at the national and international level may prove to be a much more difficult issue to resolve. As more countries turn to the auctioning of frequencies at the national level such as the US. and New Zealand, the pressure for international auctioning of frequencies will undoubtedly increase.

4.2.4 Earth Station Reliability

The operation of an earth station hinges on many factors such as reliable power (the most frequent source of failure), sufficient margin against rainfall, maintenance and repair of components, and operational techniques to minimize the effects of sunspot actively and direct sun interference. The highest reliability can be achieved by site diversity. In the most reliable form this means a link to another earth station antenna which is in turn interconnected to a back-up satellite. Several new factors on the reliability and system availability side of satellite service are brought into play by the advent of digital communications. Digital communications can aid system availability, increase service capacity, enhance quality and expand data throughout per Megahertz of available frequency. Yet this digital advantage can disappear without the maintenance of perfect synchronization which is critical for new digital services. With everything pegged to bursts of bits precise timed to the millisecond, there is increased dangers of communications failure due to "timing" problems. There are synchronization "dangers" lurking everywhere. Satellites in inclined orbit can have transmissions paths that vary 20 millisecond or more in length. Downlink transmissions traveling through heavy rain can be bent and thus change the transmission time by several milliseconds. If that is not enough some digitally compressed signals such as a videoconference transmission can be subject to a built-in 80 milliseconds or more delay in addition to the 270 millisecond of delay already built into a satellite transmission.

The solution to digital synchronization is usually storage and buffering to allow a computer to store "out of synch" information until it can be reinstated in the right format in the right sequence. Buffering and storage can also be required in fiber optic cable transmissions to cope with transmission jitters and to handle digitally compressed video. In general, however, much more buffer and storage capability (by an order of magnitude) is required for satellite systems. Most manufacturers of digital modems for satellites will have built-in storage and buffering for at least 20 milliseconds. You may wish to equip for an increased capacity of 50 milliseconds or more for safety margin, or at least insure that upgrades are possible.

Finally, there are other solutions to the problem of synchronization of digital transmission. One approach is to have an on-board clock of exceptional accuracy such as a cesium clock. This satellite-based clock can then serve as a system regulator and reduce the need for storage and buffering.

4.2.5 Recap Key points about Antennas and Frequencies

Before going on to the details of how antennas work it is important to recap the basic points about the relationship between earth station antennas and frequencies. Basic rules to remember are the following:-

At frequencies in the range of Ultra High Frequency and below, (Typically 3,0 GHz and below) antennas can be relatively simple, e.g. omni-antennas, di-poles and parabolic antennas whose radius of curvature can be fairly proximate. For frequencies in the microwave or Super High Frequency (SHF) range the antennas design must be increasingly more stringent or precise. The precision of SHF parabolic antennas must be very precise in order to radiate the right frequencies in the right direction as well as to minimize unwanted side-lobe emissions in the directions which could cause interference to other satellites.

The most effective way to reduce the cost of earth station antennas operating in the microwave region is to build very high power satellites with directional multi-beam antennas which can blast through the ionosphere and the atmosphere even when rain storm are bending the signal miles off-course and even when some earth station antennas have a marginal look-angle of, say, 5 to 10 degrees rather than a "comfortable" 40 or 50 degree look-angle.

In short, today, technology inversion is working. In some cases a dollar spent on-board the satellite can produce a ten-dollar savings or more in the ground segment assuming there are enough antennas involved. Of course it can also work the other way around if a system is badly engineered, or if the number of antennas in the satellite networks are small.

Recognize this. There are few "million dollar secrets" hiding out there. Adding clocks or program-tracking to earth stations could save millions of dollars in space segment charges by allowing the use of extended life satellites. Phased array antennas could also produce big dividends in the late 1990's. Even so, there are few miraculous innovations in the space business. Most cost savings in space communications services come from good basic engineering and effective network planning.

4.3 THE NUTS AND BOLTS OF AN EARTH STATION ANTENNA

The essential elements in an earth station antenna is provided in Figure 14. Let's walk through the entire process starting with the incoming telecommunications by a terrestrial route—let's say by coaxial cable. Multiplex or Mux equipment converts the signal into baseband frequencies and onto appropriate carrier waves that are to be routed to a particular earth station antenna which is contained in the signaling information.

Modulation comes next. If the transmission is to be digital, the modulation technique used will likely be Pulse Code Modulation (PCM); if on the other hand it is to be analog, then the modulation technique to be used will likely be Frequency Modulation and would thus work with Frequency Division Multiple Access (FDMS) multiplexing.

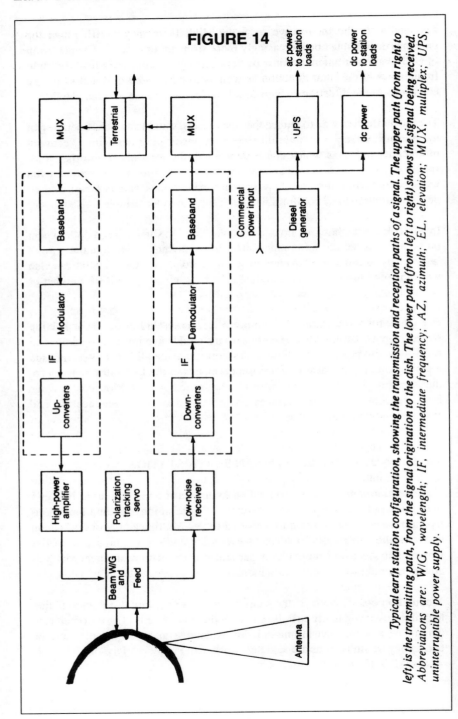

FIGURE 14

Typical earth station configuration, showing the transmission and reception paths of a signal. The upper path (from right to left) is the transmitting path, from the signal origination to the dish. The lower path (from left to right) shows the signal being received. Abbreviations are: W/G, wavelength; IF, intermediate frequency; AZ, azimuth; EL, elevation; MUX, multiplex; UPS, uninterruptible power supply.

The next step is to convert the baseband or Intermediate Frequency into the microwave or Super High Frequency (SHF) that will be used for the earth- to space part of the link. This task is performed by what are called the "upconverters.' In the case of C-band communications the signal would be converted to somewhere in the 5,900 to 6,400 MHz range.

The signal must next be routed on to a high power amplifier (HPA) in case of a large earth station or perhaps it will be only a solid-state amplifier in the case of a smaller and less expensive antennas. This process readies the signal for its journey to the satellite which is motionlessly (or almost so) awaiting the transmission nearly a tenth of the way to the moon if the transmission is to a geosynchronous satellite. Since the distance to be traveled is so great and path loss so huge (i.e. 200 dB in path loss), it is important that maximum amplification be achieved, but at the same time distortion of the original signal must also be held to a minimum. The amplified signal is then sent through a very high quality, low-noise beam waveguide directly into the feed system of the antenna and irradiated into space.

It sounds like a complicated and time-consuming exercises, but it isn't. The entire electronic transaction takes place in a few milli-seconds. The trip to the satellite even at the speed of light requires considerably more time — about 125 milli-seconds in fact. If you think of a second being equivalent to saying "One Mississippi," then think of 125 milliseconds being equal to saying "One... very quickly."

When the signal reaches the satellite antennas it completes a similar set of transactions as on the ground except the frequencies are translated down from the 6 GHz range to the 4 GHz range. As you recall the lower frequency travels more easily through the atmosphere during rain storms upon re-entry into the earth environment. The re-amplified signal now in the 4 GHz range is, within milli-seconds of its arrival being transmitted toward the designated earth station in the case of telephone call or in the case of say television is being beamed to what could be a very large number of earth stations indeed. If one takes the extreme case of a global television event such as the Olympics, the Miss Universe contest, or a Live-Aid show, the telecast might go to one or more transit antennas which would in turn relay the signal to another satellite in a different ocean region which might re-transmit it to dozens of other new location or conceivably relay the signal again to yet another region. In such global television events, the use of 5, 7 or even 9 different satellites is actually fairly common. Double or even triple-hop connections between the Atlantic, Pacific and Indian ocean regions are almost routine.

4.4 TRACKING A SATELLITE TELEPHONE CALL

To understand how a satellite connection works lets track through a complete international telephone call is routed on a geosynchronous satellite system. Let's assume that a son in Chicago, Illinois, is trying to call his parents in Rome, Italy. The call would probably come by fiber optic cable or microwave relay from Chicago into the Roaring Creek, Pennsylvania International Earth station. This incoming terrestrial call would then go out on the Roaring Creek Atlantic Primary earth station antenna on a 6 GHz uplink. The satellite would process the call and downlink it at 4 GHz into the Fucino earth station near Rome. At this stage, another 125 milli-second into the transmission the process which had initially worked at Roaring Creek begins to work in reverse at the Fucino earth station antenna.

The signal is bounced off the antenna reflector and into the feed and beam waveguide system and from there directly into the low-noise receiver. This filters and amplifies the signal. Next, the down converters translate the frequencies to the IF or Intermediate Band. The signal is next demodulated and put into baseband carriers whereby multiplex equipment formats the signal to go perhaps by terrestrial microwave relay or fiber optic cable to Rome. The signal will probably be switched a couple of times in the terrestrial exchange system before it arrives at the designated apartment.

One could be perverse and say that Mrs. Cassini and her husband Luigi were out shopping the on the Via Veneto, and thus the dozens of electronic transactions that had occurred between Chicago and Rome in an incredibly short elapse time of about 300 milli-seconds were all for naught.

For the sake of a complete explanation let's assume the Cassinis are home, however, and that the call went through. In another 300 to 400 milli-seconds a complete two-way telephone circuit is established between Rome and Chicago and Chicago to Rome. The return link is the mirror image of the incoming call.

First the "return" call is routed through the telephone network and the local exchange in Rome and then sent by microwave or fiber to the Fucino earth station. The signal is next multiplexed and converted to baseband frequencies within the carrier assigned to Italy — US. traffic. Then it is on to the upconverters to translate the frequencies to 6 GHz.

The final stage prior to transmission up to the satellite is a run through the high-power amplifier (HPA) to ensure that the signal can be "heard" by the satellite. Next, the signal is sent by waveguide out of the feed, bounced off the antenna reflector and from there the modulated voice races 22,238 miles into space before the waiting satellite intercepts Mrs. Cassini very faint message and begins to process it, translate its frequencies to the 4 GHz range and amplifies it prior to relaying the message back to Roaring Creek, Pennsylvania and then on to terrestrial networks in the US.

A complete round trip circuit can involve more than a dozen switches and a journey almost equivalent to halfway to the moon. Even so the whole transaction can take place in less than a second.

The consumer is little aware of the complexity of the transmission since everything happens in transaction times measured in milliseconds. Detailed planning ensures that carriers of adequate size are available to handle traffic demand. Very large carriers are available 24 hours-a-day on the heaviest routes such as the United States to the United Kingdom; Japan to Korea; or Spain to Mexico.

Medium to small size carriers are used on less busy streams such as Hong Kong to Indonesia, while other very thin routes such as Venezuela to Burkina Faso may be on demand or on a single channel per carrier basis. The use of FM carriers to transmit overseas calls is, however, quickly changing. Digital service is replacing analog service especially among developed countries. In the case of digital transmission efficiencies do not change greatly as in the case of FM where, in effect, economies of scale apply.

4.5 THE ANALOG TO DIGITAL TRANSITION - KEY TO THE 1990'S

Dedicated carriers using Frequency Modulated analog techniques although now still in use on the international satellite system (INTELSAT) will soon start to vanish from the scene. The new digital revolution is changing all this. Analog telephone service regardless of whether the terrestrial or space communications systems can be terribly inefficient. Analog telephone systems have traditionally worked by establishing a full -time two-way channel or circuit from speaker-to-speaker.

There are some obvious flaws in such a system. If you wish to use the telephone to transmit data in a single direction obviously half the capacity is wasted. Even in a telephone conversation each of the channels is used only about forty percent of the time. As noted earlier in Chapter Two this is because forty percent of the time one speaker speaks, forty percent the other speaks, and twenty percent of the time is represented by gaps, pauses and reaction times.

With new digital communications systems using Pulse Code Modulation (PCM) and Time Division Multiple Access multiplexing one can start to create telecommunications links that are much more efficient. In particular you can create demand assigned capacity that grows or shrinks with actual traffic demand. Since digital communications are based upon the coding of information in tiny increments measured in milli-seconds, it is possible to create links or for that matter to discontinue them hundreds of times a second. This allows total flexibility. If one needs, full-time one-way data channel this can be instantly provided.

Voice channels are available instantly when you are speaking through a technique known generally as Digital Speech Interpolation (DSI). As explained earlier the digital channel "appears" when you talk and disappears when you are not talking. This technique alone more than doubles available capacity.

Furthermore, as explained under the section on frequency usage, efficient low-rate encoding (LRE) techniques already include 16 and 32 kilobit per second voice service. In the future, that is in the early twenty-first century, 9.6 and even 8 kilobit per second voice may also be introduced to make voice communications even more efficient. In terrestrial cellular radio systems there are already 4.8 kilobit/second voice channels. The key to all this new efficiency is "intelligence" in the network. Computer processing is increasingly substituting for telecommunications throughput.

The new characteristics of digital satellite communications are constant sampling at a rate much faster than machine gun fire. Low rate encoding and digital speech will thus combine to produce enormous efficiency gains of ten times or even more. In a few years time there will be a shrinking number of dedicated full-time voice circuits using a preassigned analog carrier channel. Few will be able; to sustain such inefficient communications systems. Even in developing countries, the trend to depreciate old analog equipment and replace it with new digital equipment will become quite pronounced within the next five to ten years.

If the digital revolution is coming, then how much will the basic diagram of an earth station antennas as given in Figure 14 change in the new few years? The answer fortunately is very little. The modulation equipment will of course be changed from analog to digital. Typically, this will include PCM encoders and perhaps Time Division Multiple Access (TDMA) or Code Division Multiple Access (CDMA) multiplexing equipment. This equipment will be highly dependent upon a super precise clock and in addition will need to also be connected to buffer and storage equipment that will be used upon demand to reestablish synchronization when there are occasional discontinuities in the space transmissions. Most likely the storage and buffering will largely be needed during major rain storms wherein rain attenuation distorts the transmission signal and bend the path sufficiently to make it long and thus make the transmission time loner. As noted earlier excursions of the satellite along the east-west orbit and even more so north-south "wobble" or excursions off the equatorial plane by inclined orbit satellites. Inclined orbit satellites are yet another reason for requiring storage and buffer equipment.

4.6 GLOBAL STANDARDS AND ISDN

The biggest change in terms of overall digital transmission efficiency, however, will likely come at the interface between the earth station antenna and the terrestrial telecommunications network. Today when the ground communications networks intersect with the space segment at the earth station antenna the problems of compatibility have gone from bad to worse. In truth with trends in fiber optic transmission and new multiplexing systems such as Wave Division Multiplexing, the interconnect of space and ground systems have become not only more expensive but more complex and awkward as well. To put not too fine a point on it, it is a "Dog's Breakfast".

Telephone services go through so many switches, modulation techniques, frequency conversions, and multiples operations that it seems at times almost miraculous that the signal goes through at all. In other cases such as television, the situation can be even more confusing in the case of international relays. This is because extra steps to convert television standards are also involved to get the signal into either SECAM, PAL, NTSC, or B-MAC; C-MAC; D-MAC or D-2MAC formats. There are hopes that the new digital service and in particular the Integrated Service Digital Network (ISDN) might change all this. The idea, in theory, is that there can be a uniform global set of digital standards that will let all communications systems worldwide, terrestrial and satellite, be compatible with one another.

So far as ISDN standard making at the international level has taken place within the International Telecommunications Union (ITU), the International Standards Organization (ISO), and the International Electro-Technical Commission. Various national and regional standards-making units beyond the international bodies are also involved like the newly created European Technical Standards Institute, (ETSI). In the US. alone there is the TI body which is part of the American National Standards Institute, the IEEE standards committees and the National ISDN Users Forum coordinated by the National Institute of Standards and Technology (the former National Bureau of Standards). Within TI there are dozens of committees such as TIXI and TISI which worry only about ISDN and Broadband ISDN. It doesn't even stop there IBM has its SNA coordination groups. And the beat goes on. The bottom line is that global harmonization seems somewhere between difficult and hopeless. Satellites vs. fiber optic cable interests, vertical vs. horizontal network architectures and data vs. voice are only among the obvious problems. US. video applications are clearly at odds with the rest of the world and finding a global standard for High Definition Television (HDTV) is like pursuing the Holy Grail. Put international trade and commercial interests on top and the standards issue becomes even bleaker.

There are at least a few points which are important to know about ISDN at this stage as follows:

ISDN Standards: The most important two ISDN standards are: (a) quality of performance expressed as a Bit Error Rate of 10^{-7}; and (b) reliability or system availability equivalent to 99.98 percent. This means quality that allows less than one bit error for every 1,000 pages of text sent. It also means system availability so great that the network can be out of service only 90 minutes or less during an entire year! These are obviously tough standards to meet, but satellites can and have done it.

ISDN Rates: The fundamental ISDN rate called the Basic Rate Interface (BRI) is 144 kilobits per second. This comprises two B channels (or bearer channels of 64 kilobits per second) plus a data or D channel of 16 kilobit per second. In short, the basic ISDN unit is 2B+D channels or 144 kilobits per second. The higher level ISDN service called the Primary Rate Interface (PRI) has both an American and European version. The American primary rate is 23B+D which is 1.544 megabit per second (The American T-1) while 31 B+D or 2.048 megabit per second represents the European T-l. Digital broadband ISDN will operate at the SONET standard in the US. at 155 megabit per second and will operate at a similar speed in Europe under the Digital Synchronous Hierarchy (DSH).

ISDN service is built up on 7 successive protocol layers. The first three layers are to establish the "machinery" to provide ISDN service. It is critical that these first 3 layers as defined be fully compatible with satellite transmission. Efforts are currently under way to insure such full compatibility is achieved.

Open Vs. Closed Networks: ISDN is intended to be an "open" network with "universal ports" wherein everyone everywhere can simply plug-in. Business users with closed private data networks are in many cases intending to keep their networks "as is" but to design special interfaces to connect to the public ISDN network. Satellite user organizations are thus moving to meet ISDN standards but to offer both "open" and "closed" options.

Although all the ISDN digital systems now being defined and implemented are synchronous systems. Packet data systems that might be sent in the so-called

Asynchronous Transfer Mode (ATM) are under study for very fast telecommunications systems that might operate at super high speeds of perhaps ten gigabits per second or even higher in the next decade. Most satellite systems today are geosynchronous systems. These are broad-band and robust enough to meet ISDN standards, but the latency or delay in the transmission path are a problem. The new LEO and MEO satellite systems are much better to meet latency requirements but their lack of bandwidth at least for the 2 GHz and below systems will have difficulty with the quality and signaling standards.

4.7 CONCLUSION

Earth stations design, procurement, installation and operation are not easy propositions. The points to bear in mind are as follows:

- Earth station costs are going to keep going down; VSAT, personal earth stations, microterminals for data distribution, USAT, hand-held mobile transceivers, and low-cost video terminals are exciting new options that will be discussed in detail in the next chapter.

- Don't try to over optimize and cut the margins in your earth stations network

- Examine closely digital technology and especially digital circuit multiplication equipment.

- Keep an eye on new development such as ISDN and ATM. Their importance will grow in coming years.

- Also if you are in video distribution you may particularly wish to explore inclined orbit operation, sometime referred to as "wobble sats". (New low-cost tracking systems are now on the market).

- Stay up-to-date on ISDN and new digital TV standards.

▆▆▆ CHAPTER 5 ▆▆▆

VSATs, USATs, AND HAND-HELDS: AN AMAZING GROWTH MARKET

> **"The hottest part of the satellite market are VSATs and the new micro-terminals."**

Simon Bull, Consultant
Comsys Ltd, London, U.K.

5.0 INTRODUCTION

We now know that a VSAT is a small, inexpensive, high performance digital earth station which is certainly very popular in the USA and becoming more so all over the world, but is it really worth all the hoopla? The answer in a word is yes!

VSATs are today changing the dimensions and the scope of satellite communications in the United States and around the world. Let's just do a quick inventory of VSATs in the United States. The Walmart and Walgreen VSAT networks now include over 3000 VSATs. Southland or as it is commonly known Seven-Eleven Stores has climbed above 1500. The K-Mart network is well over 2000 and growing strong. Target discount stores now have 500, while the American Farm Bureau networks stands at over 3000. Dozens of other retailers, service organizations, and transportation, power and energy related corporations also have VSAT networks frequently with more than a thousand small antennas in their network. The current US. record holder is the Chrysler Motor Corporation with over 6000 VSATs on line.

Today VSATs in America are not a fad or fluke but a major telecommunications trend line. Altogether there were some 116,800 VSATs in operation in the United States as of the start of 1995. There were also nearly 50,000 VSATs in operation in the other parts of the world.

The VSAT revolution which started with news agencies establishing networks in the US. have now spread around the world. There are at least 6000 VSATs in India, over 4000 in Indonesia, over 4000 in Australia, over 2000 in Brazil and many thousands more in Mexico, Canada, the United Kingdom, Italy, France and China among others. By percentage the breakdown of the estimated 165,000 VSATs around the world today is as follows: (a) The United States—71%, (b) Asia-Pacific—9%; (c) Europe—10%, (d) Latin America—7%, and Africa and Middle East-2%. What is particularly different about VSATs is that in the past antennas were ordered in lots of 1 to 100, but VSATs are frequently ordered and deployed in lots of 500 to 10,000. The scope of the impact of a full scale VSAT network can be huge.

In short, the spreading applications of Very Small Aperture Terminal (VSAT) systems could totally redefine the character of satellite communications. Today VSATs are many things that they were not a few years ago. They are now smaller, cheaper, and have much higher throughput. Most importantly, there will be hundreds of thousands of them by the end of the decade.

VSATs are now found in scores of countries — developed and developing alike. With rapid technical innovations, development of new services and increasing deregulation of satellite services, it is difficult to monitor the phenomenal growth of VSAT terminals at the national, regional, and international levels. In the early 1980's the growth of small earth stations remained relatively moderate. Today, however, there appears to be a new stage of growth with over 30,000 terminals now on back-order from US. manufacturers alone. The average number of order of VSATs worldwide per year for the last five years is just over 40,000.

It all started a decade ago. The first phase was one-way low to medium rate data distribution to spread spectrum micro antennas (75 centimeters in diameter) helped to fuel the first wave of growth of VSATs in the early 1980's. Today, both the international and domestic VSAT markets seem to be evolving away from one-way service towards more sophisticated interactive applications, able to support voice plus data services.

These changes served to make the VSAT industry more vital, but has also tended to complicate the issues of just what a VSAT is, what a VSAT does, and where does conventional satellite communications begin and end. In short, is a VSAT a "bypass" technique that allows a few private business networks greater flexibility, or is it a bow wave of a whole new type of telecommunications architecture which allow users much more direct and versatile communications?

Several fundamentals represent useful starting points. VSATs are not necessarily receive-only "terminal" facilities. Many are interactive and can send or receive voice and data service to any point in a network. Some are larger (2.4 to 3.5 diameter antennas). VSATs with higher power can even support 2-way video conferences at the T-1 rate (e.g. 2.048 Mbit/s) in Europe of 1.544 Mbit /s in the US, Canada and Japan.). These T-1 rate systems are sometimes referred to as TSATs. VSAT and TSAT operations are being upgraded in many instances in order to meet ISDN standards of 99.98% system availability and Bit Error Rates (BER) of 10^{-7}. In short, VSATs are becoming more mainstream.

VSATs, regardless of whether they are for international, regional or domestic service are "driven" in performance and cost by the configuration and performance of the hub stations. The recent trend toward shared hubs, automated production techniques and other cost reduction measures have undoubtedly contributed to the recent growth in the VSAT industry.

Although today's systems are based upon a star configuration with an expensive hub station in the middle, the new Mesh configuration that allows direct interconnection of everybody to everybody else within a large VSAT network is currently under intensive development in the US. and Europe, especially with the NASA ACTS test and demonstration project. If this proves economic and technically feasible it will open up a whole new range of VSAT applications that is far wider than today's VSAT hub networks. The long term prospects for VSATs may be very bright indeed as it penetrates the mainstream telecommunications market to an ever greater degree.

The success of VSAT earth stations, in fact, now promises to spawn two new extensions of this technology. One new trend is the move to create micro-terminals that are considerable smaller than VSATs. These antennas have an aperture size of 50 centimeters. These can be of conventional parabolic shape or in the higher frequencies they can also easily be flat or phased array antennas as well. These USATs become possible with the creation of super powered thin "pencil" beams connected by on-board processors on the satellite.

The other trend is created by the development of the low and medium earth orbit satellites. The relatively high power of low earth orbit satellites and their superior look angles allow, satellite planners to think not only of Ultra Small Aperture Terminals (USAT)s but even hand-held transceivers that use simple omni antennas or patch or phased array antennas.

Regards of whether one speaks of USATs or hand-held transceivers, the overall scope of change is the same. We are still moving toward technology inversion which means higher powered and more sophisticated satellites which work to simple, cheaper and increasingly portable antennas. This is not to say that the future will be only micro-terminals. Rather, it is to suggest that we will see an expanded range of antennas from large, high-density earth stations carrying gigabits of data a second, down to very small micro-terminals. Finally the new HALE platforms will also spur the micro-terminal revolution as well.

5.1 THE VARIOUS USES OF VSATs

VSATs, however, are not a panacea. They offer specific solutions for specific communications needs. They fill a niche in the communication world that previously had not been met by space or terrestrial communications networks. VSATs are uniquely qualified for the following roles:

(a) Broadcast distribution of data to a very large number of widely distributed points. Most of the major news services such as Associated Press (AP), United Press International (UPI), Reuters and Agence France Press International employ VSATs to distribute news and even high quality still photographs.

(b) A low to medium rate interactive data network serving a whole area, which involves scores of locations and hundreds of pathways available upon demand. Such private business networks are today the "cause" of VSAT growth. (In a recent survey reference of satellite communications in the US, close to 37% of all satellite traffic in this market was private network traffic and this was projected to grow to over 50% in the next 10 years).

(c) Voice and data links to isolated locations (e.g. islands, oil drilling rigs, commercial sites n jungles, deserts, etc.) These are often linked into the private business network as described above.

(d) Applications which require bypassing the regular telecommunications system for reasons of security, economy or sophisticated applications that cannot be sustained in the regular network. The military authorities in an increasing number of countries have found VSAT networks (from UHF to X-Band) to be key to their communications needs.

5.2 THE THREE TYPES OF VSATs

There are essentially 3 types of VSATs: one way spread spectrum terminals (typically with antenna diameters of 0.7 to 1.0 m), interactive spread spectrum systems (typically 1.2 to 2.4 m in diameter), and BPSK interactive systems capable of handling 64 Kbits per second up to T-1 carriers (i.e. 1.5 to 2 Mbits per second) digital carriers for transmit and receive. These antennas are typically 1.8 to 3.5m in diameter. The smaller apertures are more frequently found in the higher powered domestic systems while the larger apertures are found in the international systems like INTELSAT.

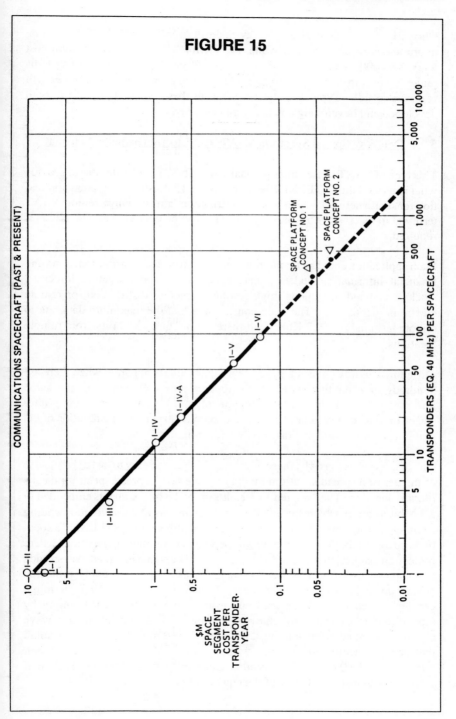

FIGURE 15

Currently, the smallest size one way micro-terminals cost $2,500 (including microprocessor and printer), while the intermediate size 2-way system cost $6,000-$8,000. The largest and most sophisticated 2-way interactive systems range in costs from $10,000 to $25,000. It is expected that these prices will continue to fall. Typical applications for all these types of VSATs in both developed and developing countries are described below.

5.2.1 One Way Spread Spectrum VSATs (0.75 to 1m Diameter Antenna)

This type of VSAT is used internationally with INTELSAT satellites to provide what is termed the INTELNET service. Figure 15 illustrates a typical application for such a network which broadcasts a data stream allowing a remote VSAT terminal to receive a 9.6 Kbit/s signal. This can be upgraded to 19.2 Kbit/s if required.

Such application could be for international scientific data distribution, news and financial information updates, meteorological data and weather forecasts, stocks and bond quotations, high resolution picture distribution, or remote electronic publishing. The main constraint is that the maximum data rate to remote terminals is 19.2 Kbit/s and there is no return capability from remote locations.

The enormous advantage of this system is particularly apparent when there are hundreds or even thousands of distribution points all requiring the same information. The terminal, including the microprocessor, antenna and printer can be produced in volume at a price of about $2000 per unit. Installation of the unit is straightforward and takes only a few minutes.

One of the key reasons why these micro terminals are so easy to install is because of the use of a spread spectrum modulation method. Spread spectrum means literally just that. The "wanted" 9.6 Kbit/s or 19.2 Kbit/s signal is embedded in a 2 Mbit/s carrier which spreads the signal by means of a convolution coding technique. The spread bandwidth can be compatible with INTELNET allocations up to 9 MHz. This technique was perfected in military applications where protection against jamming had a high priority. The signal is spread over so wide a portion of the spectrum that the wanted message is effectively protected against noise, jamming or interference. This means an INTELNET1 terminal can be located in an urban area and usually avoid interference simply by repositioning the terminal to achieve screenings from the unwanted microwave signal by use of adjacent buildings, walls, etc. The terminal aperture is so small that satellite acquisition is easy without any need for precise alignment. Furthermore the 0.65 to 0.75m terminal can even be located inside an office, if there is an unobstructed view of the equatorial plane.

The major economic cost factor for any VSAT service is the large cost of the sophisticated hub station. This is true of both receive only and interactive VSAT systems. Thus, the cost analysis that follows equally applies to the following two sections as well.

Each hub station (in 1995 dollars) costs a minimum of $800,000 (US) to build and then requires several full time staff members to run. The implication of the high cost of these hub stations is that one of two conditions are necessary to make data distribution service cost effective: (a) there must be a large number of terminals in the network (i.e. in the hundreds if not thousands); or (b) a shared hub must be used to apportion costs among several users.

For example, if there is a hub station whose annual capital and operating expenses are $500,000 and there are 20 stations whose annual operating costs are $850 each, the net coverage cost is $25,850 per station, but if there are 2000 VSAT terminals the net average cost fall to $1,100. The impact of shared use hub station is even more dramatic. If a network requires only 15% of a shared hub time, then the new average cost per year per VSAT terminal in a 20 station network would be below $3900 rather than $25,850.

Even though these assumed VSAT prices (as of the start of 1995) may be reduced over time, the relative economics in favor of shared use hub earth station operation should remain unchanged. Shared hubs have been the key to INTELNET I type (1-way) VSAT service and now are becoming the way to provide interactive VSAT service as well.

Other ways to reduce costs have included more effective manufacturing techniques as well as joint venture manufacturing agreements such as that involving the Government of India and EPIC, a subsidiary of GTE (in the USA). Under this agreement several thousands of terminals have been manufactured for operation in India where the production costs are reduced by the use of local labor.

Despite these positive factors, this is a market which appears to be saturating. Beyond news agencies, financial serves and some scientific and meteorological services, there are not too many other applications suited for one way data broadcasting. Most business and government users require two way voice, interactive data, and even occasional use videoconference and high speed data. Interactive VSAT services as described in the following sections respond to some of these needs.

5.2.2 Interactive Spread Spectrum VSATs (1.2 to 2.4 Diameter Antenna)

The small spread spectrum interactive VSATs can receive 19.2 Kbit/s and transmit from remote locations at 1.2 Kbit/s or 2.4 Kbit/s. This type of VSAT has at least a 1.2 m aperture and (as at the start of 1995) typically costs around $7K to $10K (US.). The limiting factor for this service is that the return signal carries a very low data rate which is not very effective for many data applications and is certainly not suitable for voice communications. Although advanced algorithms for voice encoding have been developed for a "voice mail" type of service the quality of the voice is generally poor and individual voice characteristics are not clearly recognizable.

When the cost of such an interactive network is calculated in terms of hub station costs, remote terminal cost, and codec equipment for 2.4 Kbit/s "voice service," the average unit cost even with 200 terminals is of the order of $14,000 per interactive VAST antennas. This does not compare favorably with 2.4m aperture VSAT using Bi-Phase Shift Keying (BPSK) transmission which can provide higher performance at lower net costs.

The interactive spread spectrum VSAT terminal service is, in general, a limited extension of the basic VSAT spread spectrum data broadcast service. This service, unless there is a major technological breakthrough, has limited growth potential.

5.2.3 Interactive Digital VSATs (1.8 to 3.5m Diameter Antenna)

The area where the VSAT industry is experiencing the most growth is the interactive VSAT terminals in the 2.4 m range using BPSK digital transmission. These terminals can handle at least 64 Kbit/s and often can operate up to 2.048 Mbit/s. To achieve these higher rates, however, a larger terminal (i.e., 3.5 m diameter) and a lot more power is required.

The sending of these high data rates for small terminals is not yet optimum and is usually employed because of special constraints such as on oil drilling rig with limited space for a larger antenna. The main point is that the "achievable" as well as the "optimum" data rates for VSATs is moving steadily upward.

Today's mainstream VSAT system, which generally has an antenna diameter of 1.8 to 2.4m and uses BPSK transmission. It has become known in the US. market as the personal earth station. This is because the small size, low cost and great flexibility of these systems are analogous to the personal computer which individuals or small companies can adapt to their personal needs. The comparison is in many ways apt. Over the past decade a personal computer with comparable performance dropped in price from \$200,000 to \$2,000 a hundred fold reduction. Over the past decade (i.e.1985—94), a VSAT has experienced a similar type dramatic reduction from \$200,000 to \$7,000 for Ku-Band and about \$10,000 for C-Band.

5.3. THE GLOBAL VSAT MARKET

The dramatic drop in price coupled with more technically efficient terminal designs has led to a surge in the VSAT market. Applications that were one too expensive suddenly became affordable. The market has now caught up with the high expectations for VSATs in the last few years. General Motor's Hughes Network Systems (previously known as MA/COM Telecommunications) now indicate back orders of over 10,000 units.

A major VSAT market study identifies worldwide back orders of VSATs of over 30,000 units. It also suggests the VSAT market is clearly due for continued rapid growth. This study predicts that satellite revenues from private networks will almost triple in the next 5 to 6 years growing from \$800 million to \$2,500 million in the US market alone. This study also underscores the fact that private VSAT satellite networks are still largely a US phenomena. In the US market, 24% of C-Band transponders and 35% of all Ku-band transponders are being used with private business networks. This is projected to increase in 1996 to 40% of C-band transponder and 55% of all Ku-band transponders. By comparison, private satellite networks represent from 6% to 10% of the current Japanese and European markets and only slightly higher in Australia and Canada.

Indications are that this discrepancy will persist for some time. A target survey of US industry with hundreds of respondents has shown that 48% of US industry with 500 or more employees used VSAT private satellite networks today and that 58% intend to do so 5 years hence. Among programmers and broadcasters, where use of VSATs is already more common, usage was expected to grow from 60% to 75%.

Perhaps the most significant strategic issue to be resolved in the future is the relationship between wide band terrestrial ISDN services (e.g. fiber optics) and VSAT Private Networks especially for widely distributed networks with average links of over 150 kilometers. A European Space Agency study quotes telecommunications planners as saying that there is a 15 year window of opportunity for VSATs but that by the early 21st century wide band terrestrial telecommunications will be provided to even rural and remote areas of Europe. The fact that European VSAT sites have grown from virtually nil in the late 1980s to 16,000 sites today is an indication that deregulatory actions by the

European Commission are stimulating this market and that VSATs will indeed share the market with fiber optic cables.

Within INTELSAT there are three basic options when discussing earth station antennas. These are Time Division Multiple Access (TDMA) (up to 120 Mbit/s using large 13 to 15m diameter antennas); Intermediate Data Rate (up to 44 Mbit/s for service to 5.5 to 11m diameter antennas); and INTELSAT Business Service (IBS) (From 64 kbit/s on up to full transponders for service to 3.5m diameter antennas on up to 9m-see figures 16 and 17). The INTELNET type micro-terminal is, in effect, a option within the IBS service. These services except for INTELNET can all meet ISDN system availability requirements of 99.98% as well as BER of 10^{-7}. EUTELSAT, Satellite Multi-Service (SMS) service will likewise be able to provide ISDN-compatible service to small terminals as well. Counterpart offerings should be developed soon in the domestic environment.

In Japan, the advent of three new specialized Japanese satellite systems specifically designed to provide TV distribution and private VSAT corporate networks, is expected to stimulate a new domestic satellite market. This well be aided by the fact that NEC is one of the world's largest VSAT manufacturers. Since Japanese communications planners are concerned with ISDN quality it is expected that Japanese VSAT networks will be of the very highest quality. Today in Japan there are over 5,000 VSAT terminals and over 4 million DBS TVRO terminals of .5 to 1.2 meters in diameter.

5.4 FUTURE TRENDS IN INTERNATIONAL VSAT SERVICE.

VSATs began as a one way data distribution service. It is estimated that there are still currently many one way VSATs as there are multi-destination VSATs. It is expected that this ratio will change in time. The run-away growth in DBS terminals in Japan, Europe and the US., however, must first run its course during the 1990s. By the year 2000 there could be close to a million interactive VSATs and perhaps 30 to 40 million TVRO receivers.

The future will be guided by an increasing range of applications. These will likely be as follows:

(a) Private VSAT Business Networks: These will be the primary application in the Organization of Economic Cooperation and Development (OECD) countries of Western Europe, the US, Mexico, Canada, Australia, New Zealand and Japan. While VSAT implementation for domestic US. networks are vigorously now underway, deregulation will serve to make this increasingly popular in

FIGURE 16

Satellite Transmission Delay

Satellite Altitude (nautical miles)	Maximum One-Way Slant Range, Earth–Satellite (nautical miles)	Maximum Ground-to-Ground Time Delay (milliseconds)
1,000	2,393	29.6
2,000	3,788	46.8
4,000	6,281	77.6
6,000	8,352	103.2
8,000	10,469	129.4
10,000	12,550	155.2
12,000	14,608	180.6
14,000	16,653	205.8
16,000	18,688	231.0
18,000	20,716	256.0
19,323	22,055	272.6

Western Europe, Japan, Australia and Canada for while France, Germany and the UK are working out regional VSAT arrangements.

The most important element will perhaps be ISDN compatible VSAT networks. These networks will redefine the commercial use of satellites and serve to challenge the supremacy of fiber optic networks for future business telecommunications needs.

(b) Video Networks: video VSAT's will become smaller and cheaper. They will increasingly use phased array flat antennas. Most significantly, video and data reception could be combined in a single VSAT which may also be able to transmit data. Combined DBS and VSAT networks will most likely emerge in the 1990's. Business networks will increasingly look to videoconferencing, imaging and other broader band applications.

(c) Rural and Remote Services: New and smaller versions of INTELSAT's rural service known as VISTA can also be expected. These will be 1.8 to 3.5m diameter antenna digital terminals capable of being powered by batteries and solar arrays. Comsat, NEC, and other companies are already preparing

FIGURE 17

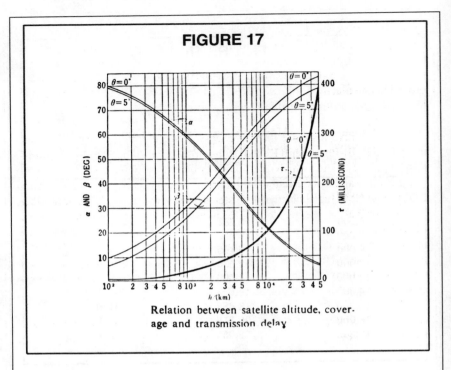

Relation between satellite altitude, coverage and transmission delay

prototypes. New competitive satellite systems such as PanAmSat, Orion, and Hispasat will offer new rural VSAT services as well.

(d) Mobile Services: Interactive low cost VSATs are the ultimate expression of the advantages of satellite communications and mobile communications units reflect the ultimate logic and efficiency of space communications over terrestrial fiber optic cable. Systems such as EUTELTRACS in Europe, Qualm's OMNITRACS in the US and INMARSAT on a global basis, point to the future of this technology. Several VSATs optimized for mobile satellite services already exist such as INMARSAT's new $5000 Standard C.

5.5 The VSAT Suppliers

The start of the VSAT revolution certainly was in the United States. When Professor Edwin Parker left the protected walls of Stanford University in order to create new type of small low cost earth station antenna, the world of satellite communications was destined to change forever. In a few short years a $2,500 receive-only data VSAT was on the market. The new Equatorial terminals became an overnight sensation. Established earth station manufacturers scurried to their labs. When they discovered that the Equatorial terminal was patent protected and that the INTELSAT INTELNET I service could be realized only through Equatorial supplied terminals a storm of complaints was created.

As a competitive response intensive work on developing low cost terminals using BPSK (Bi-Phase Shift Keyed) technology began. These terminals, unlike the Equatorial product which used CDMA spread spectrum multiplexing, could be interactive and work at higher bit rates. John Puente and Andy Werth of MA/COM, now GM Hughes Network Systems, said at the time: "This new BPSK VSAT terminal we are developing for interactive service is no more complicated than a Honda Civic or Toyota Tercel. If the Japanese can design, manufacture and ship their cars around the world and sell them for $10,000 so should we." GM Hughes Network Systems largely succeeded in this design-under-the-gun-environment. Not surprisingly so did NEC, Scientific Atlanta and a number of the other big suppliers.

Even so the number of VSAT suppliers is relatively small despite being a global market. The key organizations producing the great bulk of high quality VSAT antennas include the following:

Alcatel-Telspace
ComViaSat
GM Hughes Network Systems
NEC
Satellite Technology Management
Scientific Atlanta
Tridom (AT&T)

Although there are a number of other suppliers the above organizations are responsible for perhaps for some 90% of the VSAT orders. The VSAT market suppliers will, in fact, likely remain thin because the technology is so demanding and changes move so quickly. Further the dollar amount of the global market although growing rapidly has only recently become a billion dollar volume and the profit margins under great competition are not particularly high. None of the factors would help attract new entrants to the field.

Rapid technological innovation, ISDN compatibility, and aggressive market research will be needed in the VSAT field, if the satellite communications industry is to retain its vitality. Such rapid development, in fact, seems both essential and likely by the early 21st Century. By that time, there may well be million of VSATs will be in regular use in scores of countries around the world.

5.6 USATs—THE ULTRA SMALL APERTURE TERMINALS

New lightweight, low cost terminals are redefining what is meant by VSAT terminals. By the 21st Century it is anticipated that hand held units smaller and lighter than a briefcase (and probably costing less as well) will be in use. Even a new name is evolving to cover these mini- VSATs and this is the USAT. The unsolved issue for VSAT terminals is whether they can provide ISDN quality service directly to and from a geosynchronous satellite. Initially, the objective would be to provide ISDN quality service directly to and from a geosynchronous satellite.

Basically, the critical objective would be to provide 144 kbit/s channels. This would consist of 2B voice channels (64 kbit/s) plus a 16kbit/s data channel.

The most precise view of the future in terms of GEO based USAT terminals has been defined in the Hughes filing in 1994 for a new 12 satellite Spaceway system that would operate in the Ka-Band frequencies. There USAT terminals with a 45 cm aperture could support T-1 service and above. In time T-1 channels or higher will facilitate a videoconference channel to the famous "Dick Tracy" type radio wristwatch. These services would likely involve low earth orbit systems to support these truly micro-terminal transceivers which would weigh only a few ounces.

The very first applications of USAT will be receive only modes such as with the Japanese NSTAR DBS system. The use of the micro-terminal USATs for Satellite Collection and Data Access in rural and remote areas with store and forward messaging systems such as Orbcomm and Starsys will begin in 1995 and 1996.

5.7 HAND-HELD TERMINALS FOR MOBILE SATELLITE COMMUNICATIONS

The fascination with mobility and tetherless communications has been one of the dominant trends of telecommunications in the 1990s. Cellular telecommunications can grown from virtually nil in the mid 1980s to some 15 million cellular telephones in the US and 27 million worldwide by the start of 1995. The success of Walkmans, personal TVs, palmtop computers and faxes, and other highly mobile units will undoubtedly be paralleled in the world of satellite communications.

Once satellite planners make the necessary leap to low latency systems and LEO systems to increase effective power and look angles, the constraints to true hand-held technology can be largely removed. The Teledesic system can utilize USAT for fixed satellite services to rural villages and hand-held for mobile satellite communications services. The Motorola Iridium system, Globalstar, Odyssey, Aries Constellation, Ellipso, and INMARSAT P are also all promising the same. Even the new Brazilian low earth orbit ECO-8 systems being planned together with CTA can likely deliver a DBS service or a mobile connection to a hand-held transceiver. By the year 2000, the idea of satellite service via a hand-held unit will seem not much more unusual than a walkman CD player today.

5.8 CONCLUSION

International and domestic VSAT networks have a bright future for business networks, rural and remote services, video broadcasts, interactive TV, data networks and especially mobile satellite services.

Some observers suggest there is a 15 year window of opportunity for VSAT business networks until ISDN compatible fiber optic cable systems are installed. This view, however, suggests a static future which does not foresee the possible development of much higher performance and lower cost VSAT terminals that are indeed ISDN compatible. By 2000 to 2005, however, this technology should be both available and market competitive.

CHAPTER 6

BUYING SATELLITE NETWORKS AND SERVICES

> **The shift from Star to Mesh Networks represents a key challenge for satellite communications, but the rewards will be tremendous."**

Dean Olmstead
Hughes Aircraft Corporation

6.0 INTRODUCTION

Buying, selling or even brokering satellite capacity can be a remarkably murky process because almost everywhere you turn somebody changes the rules on you. The only thing to do is to try to start with the big picture and fight your way through to the little bothersome details.

6.1 THE BIG PICTURE

There are a number of basic postulates which generally hold true, but which still can change with circumstances and details.

Satellite service makes great sense and indeed it is often the only option available for:

(a) Maritime mobile service

(b) Aeronautical mobile service

(c) Remote and rural land mobile services

(d) National or international televisions or radio distribution or broadcast

(e) International telecommunications to developing countries

(f) Radio determination satellite service and navigation

Satellite telecommunications service is still a viable and cost effective option for overseas communications and also for service to and within countries. Nevertheless fiber optic cables are more than giving satellites a run for the money. (Key considerations here are cost, quality, reliability and diverse routing of traffic especially military traffic.)

Satellite telecommunications for national telephone in developed countries is shrinking. In fact in the US, public telephone service during the 1980's virtually disappeared off of domestic communications satellites. US. Sprint, AT&T and MCI, the three major US long distance telephone companies have all largely dumped their satellite services for regular voice services over terrestrial facilities. In Japan and Europe the trend is largely the same with a few exceptions like the German DFS-1 Kopernicus and the French Telecom satellite projects. Today's national satellite markets, really tend to have only two major components: broadcasting (radio and television) and specialized digital networks which can be used for data (financial records, inventories, order forms, etc.) for videoconferencing and digitally compressed voice and video.

6.2 THE KEY VARIABLES

One thing is clear. How, why and whether you want to buy satellite capacity varies on the basis of the following:-

- <u>Distance</u>: Local, state, national, regional, overseas (the greater the distance the better the service is for satellites and vice versa).

- <u>Traffic density</u>: Think route, T-1 carrier, partial transponder, full transponder, or multiple transponders. (Satellites are best for a variable mix, while fiber optics are best for a constant high volume).

- <u>Type of service</u>: Point-to-multipoint (broadcast video or data distribution); point-to-point dedicated telephone/data lines); multipoint-to-multipoint (business networks). (Satellites are, of course, best for various forms of multipoint service while terrestrial systems are best at point-to-point).

- <u>Bypass considerations</u>: Satellite networks which connect to the space segment directly from the customers plant provide the ultimate bypass option. This can provide quality enhancements, tariff advantages, and service flexibility. Overall this should add up to significant cost reductions.

- <u>Available terrestrial connections</u>: The more robust, modern and digitally sophisticated the terrestrial telecommunications network is, especially in terms of fiber optics cable and electronically switched exchanges, the less attractive satellite services will tend to be. Alternatively if your service requirements involve remote or ill-equipped areas you may want satellites. (Don't jump to conclusions, however, since in some cases parts of New York City there is installed telecommunications equipment that looks and works like it was a leftover from a Civil War rummage sale.

- Quality, reliability, availability and security: Depending upon special requirements in this area satellites may provide all or at lest part of the answer. Sophisticated users are now often insisting on multiple or at least parallel routing of their traffic. Tandem computer has built its whole service on having two completely independent end-to-end satellite routes in service. Specifications of your service requirements in these areas are critical in getting what you paid for and ensuring it fulfills your needs.

Cost: Cost is almost always the bottom line. Yet it is hard to compare costs because:

(a) Satellites and terrestrial communications have unequal capital equipment costs, operating costs and restoration costs.
(b) Satellites tend to allow you more bypass options.
(c) Common carriers give you fixed tariffs but most satellite systems who are selling raw capacity (or brokers who are selling capacity in smaller chunks) represent a volatile market. Prices for a transponder or a part of one are rapidly moving targets.
(d) International services are even more difficult to price and maintain. Suffice it to say the US national market for satellite or terrestrial services is the world's lowest price offering. Domestic US transponders are two to even eight times less than other locations around the world. (If you want a large overseas satellite telecommunications network, definitely, but definitely, get some heavyweight consultant support.

6.3 BUYING SATELLITE COMMUNICATIONS IN THE U.S.A

There are at least three ways to purchase capacity for satellite communications services in the US market. Typically there are fewer options in the other countries of the world. In fact there are very limited and quite precise ways to obtain capacity elsewhere. The rest of the world will be addressed in the next section but the more complicated case of the US. will be addressed first.

6.3.1 Common Carrier

To become a common carrier in the United States has in the last decade become a remarkable simple thing. If you have a going company with both an interest and a capability in telecommunications you could probably become one. The current process is to file an application with the FCC and show reasonable ability to perform. The FCC staff, not the full Commission, approve such applications. Today there are about 1200 carriers of various types and varieties in the US. This is down from the more than 1500 of a few years ago.

The key is that today US. carriers do not have to have a large investment in telecommunications facilities. They can lease, purchase or sublease facilities from larger entities and then go into business as a re-seller. Two examples demonstrate the point. One company called International Relay Inc., out of Chicago, Illinois, (now sold to another carrier) obtained FCC licenses in 1984/1985 to build a number of 7 to 11 meter INTELSAT Business Service (IBS) antennas around the US. (New York, Chicago, Atlanta, Washington, DC. etc.). They also arranged to lease through COMSAT a number of INTELSAT business digital circuits at rates ranging from 56 kilobits up to T-1 (1.5 megabits/second). The FCC not only granted them a license as an international carrier but also worked with the US. State Department and US. Commerce Department (NTIA) to get them correspondents in Germany, France, the United Kingdom and Japan.

The point is by putting in about 5 million dollars of capital and a year of effort IRI suddenly was a functioning international carrier by 1985. IRI bought their satellite capacity essentially wholesale and then retailed it at published tariffs formally placed on file at the public documents room of the FCC. The point is the tariff rates for the "new" carriers are typically less than the established big guys — sometimes up to 30% to 40% less. Of course the quality and reliability of the service can vary greatly too.

The second example of a "new" carrier is an interconnection carrier for local and long distance telecommunications services in the US. There are dozens of these entities who offer personalized and customized telecommunications services. They typically will offer a combination of satellite and terrestrial requirements as needed. Often the prime focus is not even on the design of innovative telecommunication networks.

Some of the new telecommunications carriers and evolve from a real estate development background. Their business plan is to utilize satellite communications to be the lever from which to create new business developments. The emphasis will frequently be upon customer premise equipment such as digital PABX equipment, especially to meet specialized needs such as "digitally compressed" 16 kilobit/second voice, 384 kilobit/second videoconferencing with new, improved and low cost codec equipment or say Group 4 facsimile offerings.

Long distance transmission, whether by satellite or fiber, is typically among the last thing considered rather than the first priority of such building and real-estate oriented business group who have entered the specialized communications market with little telecommunications experience and even less international exposure.

The "services" that common carriers provide to the user community are considerable. They include:

(a) Clearly understood and specified services.
(b) Flexibility to upgrade or downgrade quality, reliability or service characteristics on demand.
(c) Ability to modernize equipment as technology improves.
(d) Clearly tariffed rates which reflect a competitive market place but are still regulated by the FCC.
(e) Responsibility and accountability for performance. (For any tariffed service, your carrier is responsible. If performance is poor or too costly you can change to another carrier).

Typically if one decides to use a carrier, you start by selecting one or several competent organizations for detailed consultation. You would ask them to review your network design and total requirements to produce a price quotation. If you are a larger operation you may seek a competitive bid from a larger range of carriers.

Since common carriers must offer like services and tariffs to all their carriers, it is not at all a comparable process to seeking bids for say a piece of electronic equipment. Even so with new environment of Software Defined Networks (SDN) or Virtual Custom Networks (VCN), common carriers are beginning to look a lot more like private network providers.

Tariff offerings by common carriers especially AT&T, US Sprint and MCI have become much more "innovative". AT&T has Tariffs 12, 15, and 16, MCI has VNET, and US. Sprint has their VCN tariffs; but they all add up to highly competitive rates intended to undercut private networks.

Some of the key variables for common carriers come in such areas as service call response time, payment schedules, and "bells and whistles", that go beyond basic service. If you wish a primary rate ISDN channel with 99,98% system availability but with a super high quality bit error rate of say 10 to the minus 10th, 99% of the time, a common carrier will probably tariff it for you. This is all to say if you are seeking clear cost and exceptional technical performance information from a common carrier, you should receive a good response with little or no surprises. If you do not get a good response then there is something seriously wrong.

Tariffs and services are clearly defined and recorded with the FCC. Once the tariff is filed, accepted and recorded, there is very little room for maneuver or favoritism ever for a large preferred customer. Even so the new Virtual Customer Networks (VCN) are redefining the rules. AT&T's Tariff 12, MCI's VNET tariffs and US. Sprint's VCN tariffs all allowing much broader tariff discretion among the common carriers in ways once reserved for private networks.

Most common carriers are not wedded to a particular transmission technology and thus typically can offer all terrestrial, all satellite, or hybrid systems. The "package" might be all terrestrial voice service but with videoconferencing and data distribution networks provided via satellite. Unless you have a precise and special constraint which require a particular mode of transmission it probably makes sense to let the carrier make this decision - not you.

If the common carrier approach appeals to you, perhaps your first step should be to contact the FCC Public Document Room (through an appropriate Washington, DC. based agent) and obtain a good cross-section of publicly filed tariffs for satellite telecommunications services. These will indicate just what services are available at what price. The address and telephone number are:

> Public Document Room
> FCC
> 1919 M Street, N.W.
> Washington, DC. 20036
> Tel. No. 1 (202) 632-7000

6.3.2 Purchasing a Complete Telecommunications Systems

There are essentially two ways to purchase capacity, namely to buy a "turn key" system or to design, engineer and implement your own system with the ability to purchase satellite capacity, earth stations, terrestrial tails and other system components on the basis of best cost and technical performance. This involves planning, bidding and procuring, integrating, testing, fine tuning, upgrading and managing the communications systems yourself.

A basic word of advice. "Don't". That is don't do it unless you really have the in-house interest, expertise and long-term commitment to see it through. It can be done. ARCO Oil Company designed and put together a highly effective state-of-the-art videoconference and telecommunications system. Other large organizations such as Sears, Proctor and Gamble, and insurance company consortia have also created their own satellite communications network from the ground up. There are twenty times more who will go to a common carrier or to a telecommunications consultant firm to give them a turn key system rather than trying to do it on their own. Building your own system does not necessarily save money and it involves greater financial, operational and technical risks. On the plus side it may be the only way to get exactly what you want.

To develop your own network, if it is to be a satellite based system you need to start with specifications for the satellite performance and the appropriate class or size of earth station antennas. The rest of the system should evolve from these starting points along with the services you wish to provide. A video distribution driven system to a large number of receiving stations strongly suggests that there is a need for a high power satellite with a sophisticated uplink earth station to feed the programming to a large number of low cost receive-only antennas.

A data distribution network is similar except data require a lot less power and much smaller antennas. This means the satellite power can be less and the receive terminals can be much cheaper, more compact, and also designed for a higher frequency interference environment if need be. If you desire a very sophisticated network for video, voice, fax, data and telex services, and multi-purpose services, it is much harder to generalize. If you wish to combine national, regional, and overseas service this too is quite complicated. Although new options are evolving in competition to INTELSAT, in the form of regional satellite systems like Pan AmSat and domestic systems that provide transborder service, it is still more likely than not that two or more different satellites will be needed if you wish to link into a diversity of international locations. This is simply a matter of satellite antenna coverages although regulatory issues can also be involved.

There is in fact no such thing as a "typical" example of a baseline satellite specification as well as baseline antenna specifications that might be associated with all types of applications for data, business digital services, voice, video and so on. The diversity of transponder characteristics for INTELSAT, PANAMSAT. INMARSAT, EUTELSAT and domestic US. satellite service is enormous. Transponders can be 36 MHz, 54 MHz, 72 MHz or even 240 MHz. The power level of such transponders vary from 22 to 45 dBW. The main reason for these variations are primarily tied to two factors: (a) antenna characteristics, and (b) widely differing satellite antenna coverage patterns. To plan a satellite system to your own needs it is best to review the general specifications for all existing and planned satellite systems. These can be found in several sources such as:

Silvano Payne; <u>The International Satellite Directory,</u> (Sonoma, California; Design Publishers, 1995)

John Howkins and Joseph N. Pelton, <u>Satellites International</u>, (London, UK.; MacMillan Ltd., 1987)

There are, of course, proprietary studies prepared by telecommunications consulting firms which give even more detailed satellite system performance information but these are much more expensive.

The critical issue in obtaining a complete satellite system is getting the right space segment capacity. It is key both in terms of obtaining sufficient capacity to meet current needs as well as capacity to meet expansion needs and to restore service. Furthermore, obtaining transponder capacity can also be the biggest single financial variable as well. The cost for purchased transponders in the US. satellite market has ranged to a high of $16 million on a Hughes Galaxy satellite to a bargain basement, money losing $4 million at the height of the satellite capacity glut. Recent surveys of the US. domestic satellite transponder market concludes that there are currently some 375 C-Band transponders and 100 Ku-Band transponders now in service.

In terms of FCC licensed capacity, this is expected to grow to 530 C-Band transponders and nearly 500 Ku-Band transponders by the late 1990's. Almost everyone, however, believes that significantly less capacity will, in fact, be actually deployed. Assessments by such experts as Satellite Systems Engineering, Booz Allen and Hamilton and others suggest there in fact could be a net <u>decrease</u> in C-Band capacity from 375 down to below 300 by the late-1990's and the increase in Ku-Band capacity will be to about 250 or 300 transponders tops.

This relative lack of new capacity can be attributed to conversion of traffic to fiber optics (especially voice-only traffic) a lack of profitability in satellites given prevalent market conditions, and a general slowing in overall telecommunications demand, especially in the video area. The main conclusions of importance here is that satellite capacity in relations to demand will gradually shift as follows:

<u>Time Period</u>	<u>Satellite Market Assessment</u>
1985-86	Glut of transponders - very low prices $4-$8 million per transponder.
1987-90	Market equilibrium reached - Supply and demand in balance- Prices move upward, i.e. $5-10 million per transponder.

Time Period	Satellite Market Assessment
1990-96	Shortage of satellite capacity develops in C-Band and Ku-Band - Prices move even higher, i.e. $6-12 million per transponder. (Deficit of 70 transponders projected for US.market as of the end of 1996).

If this assessment is correct it would make sense to plan your network and obtain contractual commitments for transponders or at least an option for such capacity sooner than later. Incidentally if you want to know who might be buying US. domestic satellite capacity for what purpose, the breakout is fairly simple: (a) video represents about 50% of the total market, (b) VSAT business networks of data and voice/data represents 42% of the market, (c) the remainder (8%) of the market is mainly voice, position determination and specialized services. This small special use sector is in fact declining as a percentage of the overall satellite market.

The current supply of transponders available for purchase is definitely shrinking in both the C and Ku-Bands. Further, the number of satellite system suppliers is growing smaller as a result of recent market consolidations particularly in case of Hughes Communications and GE's buy outs. Further the GE take-over of the GTE system further consolidates the satellite service suppliers as well. A competitive procurement process to shop for transponders is not normally employed because of the "thin market".

The process is perhaps most analogous to that of shopping for a very expensive automobile. You know what you basically want in style and performance. After talking to potential suppliers you then work out what optional features you will add and then agree on a mutually acceptable price with one of a limited number of suppliers.

Some very good, common sense advice on how to purchase satellite capacity (either a transponder or even a whole satellite system) is available in published form. One of the better articles is that by Elio Sion, a US. attorney specializing in this field, in an article called "The Do's and Don't of Satellite Procurement" in the Journal of Space Communications and Broadcasting, Vol. 6, No. 3, 1988. The basic advice he offers is to not go with a rigid RFP or overly restrictive technical specifications. In some cases, for instance, a 1 dB difference which might have little or no performance impact could cost millions extra. Ranges of performance, flexible coverage patterns, etc. can result in tremendous savings often with marginal impact on practical results.

6.3.3 International Purchase of Satellite Capacity

Buying space segment in the international, overseas market is considerably different than buying capacity in US. domestic markets. There are in fact several approaches that can be taken. If one is interested in a combination of US. domestic and transborder services to Canada, the Caribbean Mexico, Central America or even Venezuela or Columbia you can consider many options. These include leasing capacity from any of the US. domestic satellite systems, from the Canadian Telesat System, the Mexican Morelos System, or even leasing an INTELSAT video transponder or INTELSAT Business Service. Typically this type of off-shore transborder service capacity is available at rates quite comparable to domestic transponder capacity. Leasing or purchase of transponders for such off-shore traffic is probably the only economically viable approach since there is only not enough traffic to support a dedicated satellite system.

There also needs to be an important footnote on the international legal process required for all international satellite service. You or the selected satellite system operator must go through what is called an INTELSAT Article XIV (d) coordination for all international satellite traffic. To date fifty plus such coordinations have been completed successfully and if the traffic is relatively modest or if this is video distribution for the US. with incidental spillover receptions in other countries, then such coordination is unlikely to create any problems or delays. There is, however, the time factor. For what might be considered a typical case, the whole process can involve about one year to complete depending upon the schedule of the INTELSAT Assembly of Parties which must make the final determination. Recently there have been attempts to streamline this process, but so far only very limited cases can be handled by the Board of Governors directly.

If, on the other hand, you wish to obtain capacity for domestic and regional services you will likely have only a few options. In Europe there is the EUTELSAT system, the ASTRA system owned by SES of Luxembourg, the TELECOM 1 satellite system whose capacity for regional service is also obtained through EUTELSAT, a limited amount of capacity from PANAMSAT or ORION and then of course INTELSAT. For mobile satellite services one can obtain capacity for maritime, aeronautical or land mobile services through INMARSAT and soon Radio Determination Satellite Services (RDSS) though Geostar.

The process in Europe is essentially twofold. For INTELSAT, EUTELSAT, TELECOM 1 and/or INMARSAT one must obtain the capacity through the relevant government telecommunications entities and there is really little opportunity to negotiate. They have fixed tariffs just like the common carriers in the US. but there is in essence no competition in the prices offered to end users.

In the case of PANAMSAT, ORION and ASTRA the approach is usually to sell capacity to their user at a negotiated price.

British Telecom, in a break with usual "European practice", has in fact become the marketing agent for ASTRA capacity. PANAMSAT has tried to sell transponders but has also attempted to obtain carrier status to lease service capacity as well. Recently the United Kingdom Office of Telecommunications (OFTEL) has ordered British Telecom and Cable and Wireless (Mercury) to accept PANAMSAT as a carrier. In general capacity whether leased or purchased outright in Europe is more expensive than in the US.

Obtaining capacity in the rest of the world involves even less options. In the Middle East it boils down to ARABSAT or INTELSAT, and in Africa it is just INTELSAT. South America offers several options and this may expand in coming years. Here there is PANAMSAT, Brazilsat, INTELSAT and to an extent Morelos. Furthermore there are plans for an Andean satellite system called Condor and an Argentine satellite system as well. In many ways this may become the most competitive region (other than North America) in terms of satellite capacities.

Finally there is the Asia/Pacific region. This region because of its great geographic size is only effectively covered by a single INTELSAT satellite, but the Pacific and Indian Ocean satellites offer a lot of options. There are, however, several domestic systems which provide some degree of regional coverage. These include Australia (AUSSAT), Indonesia (PALAPA), India (INSAT) and several Japanese systems. In addition ASIASAT system provides coverage from Southwest Burma to Northwest China. Capacity on all these various satellites are typically leased or sold in transponder units. If you wish smaller units particularly a T-1 carrier or less you will probably have to deal with a broker or re-seller who "retails" the smaller units of capacity.

The number one supplier of satellite capacity worldwide remains the INTELSAT system. At the present time just over 100 transponders have been leased or sold to 35 countries, both developed and developing, for domestic services. Over 30 transponders have been leased for international video distribution. Furthermore over 100 private international business networks ranging from small to very large have been created through the INTELSAT Business Service. While INTELSAT is the mainstream supplier, new private international satellite systems like PANAMSAT, ORION, ASIASAT, and ASTRA offer a "deregulated" alternative with all of their positive — and negative factors.

6.3.4   The Satellite Broker

The third and final way to obtain satellite capacity is through a broker or larger user with the legal right and inclinations to sublease. There are essentially two types of brokers of satellite capacity. One type is a "specialized" broker that obtains capacity which is then used to offer services to a particular and usually quite exclusive or specialized market. A broker of this type might sell time to send out video political ads, or for marketing and inventory information to car dealers, or provide a remote site auctioning service, or even health and education services. These brokers essentially see themselves as being part of a particular service industry whether it be politics, automotive sales, auctioning or education.

The communications channel is often only a part of the overall service. Videoconference rooms, on-site auctioneers, pre-event publicity, and even scripting of presentation and special video effects are often available from these organizations. It is typically the value-added services offered by these "specialized" brokers that makes them highly useful to their clients. Further, it is the special expertise and targeted services that provide the profits, not the satellite capacity per se.

The second type of broker is a wholesaler who obtains large chunks of capacity either on the basis of reselling it in smaller chunks of capacity, smaller chunks of time or both. This is most common for video applications particularly for international use. Capacity can usually be obtained for video in 30-minute increments, while voice or data channels are typically required for longer periods, typically one year. Guides to organizations that broker satellite capacity can be found in several easy-to-find reference materials. These include the International Satellite Directory (Design Publishers), the Satellite Yearbook (Baylin Publications or through the Internet by telnetting to Satnews. Com (Design Publishers).

There remains the "broker "who often can and will provide the best price of all. This is a carrier or user who has over-subscribed and has excess satellite capacity to sell. Equatorial obtained 9 transponders in the mid-1980's for data distribution services which was about twice the amount required. Subleases usually obtained good prices in this oversupply market. Citicorp once over purchased transponder capacity and then resold it. In this case Citicorp was fortunate to get their money back. Anyway before obtaining spot time or purchasing say a T-1 carrier from a broker, it is wise to survey all the larger users of satellite capacity to see if they have spare capacity to sublease.

In the US. there is little restriction with regard to such subleasing arrangements, except to ensure that transmit earth stations (particularly smaller ones) are not interfering with other satellites or other carriers on the same satellite. Elsewhere there are more strict regulations. Sometimes, it is necessary to become a formal partner to a consortium of users to qualify to use capacity initially approved for another user. The advent of ASIASAT, ASTRA and PANAMSAT (PAS) has, however, tended to relax these restrictions in time.

6.4 CONCLUSION

The way to purchase or lease capacity varies greatly around the world. Shifting patterns of usage suggest that the most cost effective and logical areas for obtaining fixed satellite capacity is in prime areas: video distribution or VSAT business network for digital services. Voice service for domestic networks is quickly shrinking as a satellite service. This is the result of rapid conversion to fiber optics service particularly in the US. Internationally, of course, satellites are still heavily used for all applications.

The first step in purchasing satellite capacity is, of course, to clearly define what your needs are in terms of capacity, security, reliability, quality, restoration arrangements and prices. You need to know what is and isn't acceptable to you. Next analyze the basic options to make sure that you wish to use satellites for all your requirements, for a share of the total, or none at all.

When you feel you have a clear idea of your needs you may wish to retain a consultant to verify your finding and to map out your strategy. Eventually, you have three options on which to process: (a) obtain the capacity from a common carrier, (b) purchase the satellite capacity from a satellite system selling transponders (and restoration capability if available), and (c) obtain the capacity either from a broker of satellite capacity or a large user with spare transponders. In only one case in a thousand do you want to buy an entire satellite system. In this case professional help is strongly recommended. The various sources materials cited in this chapter and re-listed in the selected bibliography immediately following should be of considerable use in understanding the practical aspects of an actual procurement of satellite capacity.

CHAPTER 7

SATELLITES AND VOICE COMMUNICATIONS

> "With the introduction of satellites and their attendant earth stations, the US. moved from a telecommunications system which linked a few nations bilaterally to a situation involving global dispersion of information through outer space. Instead of a link to London, Paris and Tokyo, satellites beam from anywhere to everywhere. By advancing the state of the art the US. has achieved instant global involvement."

Professor Oswald Ganley
Harvard University

7.0 INTRODUCTION

There are many different ways in which international and domestic communications are dissimilar. None is more striking, however than in the area of voice communications. Some 75 percent of the revenues from international communications comes from plain old telephone service (POTS). For most domestic satellite services, at least in developed countries, however, over 90 percent of all revenues comes from <u>non-voice</u> services, and often predominantly from video services. Thus, national and international satellite communications systems, from a service mix point of view are almost perfectly asymmetrical — like in day versus night.

7.1 WHY THE DIFFERENCE BETWEEN INTERNATIONAL AND DOMESTIC SATELLITE SERVICES

How could such a large difference exist when the technologies and the capabilities are so similar? Well, there are in fact several good reasons why this is so:

(a) <u>Existing Terrestrial Networks</u>

In the US., Europe, Japan, Canada, Australia and other countries with a developed economy there are well developed terrestrial telecommunications systems already in existence at least in all the major cities and in their major trunk line interconnections between key cities. Only rural and remote areas are under served. The telecommunications planners in most of these countries tend to think in terms of terrestrial networks rather than satellites for the expansion of infrastructure they have already built. Satellites are thus usually seen as a supplementary tool to be used for back up, overflow, or for special cases where terrestrial technology cannot go or cannot go economically say into Alaska, the Canadian Yukon or the Australian Outback.

(b) Natural Advantages: Satellite Based Global Telecommunications Coverage

It is no accident that satellites today provide two-thirds of all overseas telecommunications coverage and will retain a significant percentage into the 21st century. Fiber optic cables like conventional submarine cables before will be largely concentrated on heavy routes between North America and Europe; North America and Japan; and a few other routes like Portugal to Brazil; US to the Caribbean and Venezuela; or Japan to Southeast Asia and Australia. Of the over 2100 full-time and demand access routes that exist on the INTELSAT system today the vast majority cannot be served by cable and even when they could be served, the complicated routing arrangements would be troublesome and often expensive. Satellites are usually more cost effective for routes exceeding 1000 miles. Satellites, in particular, are a flexible, quick, and cost-efficient way for many countries to establish overseas communications. This is particularly true if you are: (a) an isolated or remote country; (b) if you are land-locked and thus well away from a submarine cable head end; or (c) if you are interested in a full range of telecommunications services including wideband television.

In addition, many developed countries like to avoid putting all their eggs in a single basket and tend to split their traffic fairly evenly between the ubiquitous satellite and submarine cable facilities. To be specific, take the case of The Netherlands. It currently has 59 overseas correspondents with direct communications. Of these 45 are exclusively on satellite links because a terrestrial option is not available or not cost effective. Only the remaining 14 routes are using submarine cable and these are split with satellites for the purpose of diverse routing. In short, satellites have many unique and cost-efficient advantages that apply to international overseas service that does not apply to national based telephone service.

(c) Bias Against Satellite Telephone Service

At a recent assemblage of the elite of the satellite communications industry known as the **Satellite Communications Users Conference (SCUC)**, a top executive of GE Americom, one of the largest operators of a national satellite system was asked about the future provision of satellite voice service in the United States. His answer, a verbatim quote, was: "The war is over".

He went on to explain that video, data, and specialized networks were still important US. national markets to be served, but that his company did not see a viable future telephone voice business on US. national satellite systems in the age of ISDN.

There is no unanimous view on this subject. but certain trend lines, however, are clear. In fact they are almost impossible to avoid. The FCC several years ago approved 25 orbital assignments for domestic satellite service. It even imposed two degree orbital spacing to accommodate all of these new and replacement satellites. Today, some four years later only a handful of the 25 satellites are "real", while the remaining twenty have been postponed or shelved.

Those US. companies who are or were offering satellite voice services are now providing less or no such service, from AT&T to GTE to American Satellite to Western Union, to MCI. Some such as Hughes Communications had never been in the telephone market. Many satellite executives look with concern at the latest attempts to create standards for ISDN and note the stringent requirements for digital telephone services in terms of system availability, bit error rates, and hot restarts.

Often they tend to throw up their hands and begin pursuing other services. Others who created the Satellite Coalition on ISDN have decided to stand their ground and fight the efforts aimed at pronouncing that fiber optic is the ultimate answer to all telecommunications problems in the US. and around the globe. So far, however, those proposing an all-fiber optic cable seem to be winning.

7.2 SATELLITES AND ISDN SERVICES

Fortunately for the satellite industry there are a number of those made of sterner stuff. These industry leaders have now joined the new Satellite Coalition for ISDN and they are indeed moving to ensure that their satellites now and in the future will meet all ISDN standards for system availability and for bit error rate. They are also ensuring that satellite voice services will be available at comparable or lesser costs than fiber optic cables. Finally, they are seeking to ensure that ISDN delay standards for voice and other services are not unrealistically restrictive consistent with actual user needs. Despite these clear, strong and proactive steps, problems involving satellite and ISDN quality remain especially with regard to digital voice. A massive advertising media campaign in the US. by a carrier that has opted for an all fiber network has created considerable mischief. Even telecommunications planners who should know better have come to believe that satellite voice is "inferior" and that fiber optic cable is "better". All the satellite industry can do at this time, short of a multimillion dollar advertising campaign, is to methodically undertake the necessary technical and operational tests that provides overwhelming objective evidence to the contrary. In fact a full range of steps are needed to counteract the anti-satellite bias for voice service in the US. market and for that matter the international market as well. Specific initiatives are needed as follows:

(a) <u>Accelerated Conversion to Digital Voice</u>

High quality, low interference telecommunication service at low cost is largely dependent upon conversion to digital service for all types of satellite service, domestic, regional, and overseas. Digital speech interpolation, low rate encoding, and digital service margins can help ensure that satellite service is competitive with fiber optic cable. Nevertheless, care should be taken. These steps should be taken only when the highest quality has been established. Voice should be kept to well above ISDN standards, or business will migrate to fiber optic voice service both nationally and internationally.

(b) <u>Digital Echo Cancellers</u>

Many people associate satellite voice service with the early days of "average" quality, "high echo" service that came with Early Bird in the 1960's. In those early days satellites with their 4 kilohertz channels for voice service were still better than the 3 kilohertz submarine cable service and vastly superior to shortwave, high frequency radio telephone that had to be bounce off of the ionosphere.

Times have greatly changed, however, and today there is no reason why a satellite circuit from half way around the world shouldn't sound as if it is from next door. High performance digital echo cancellers properly installed and maintained in a country's national and international telecommunications network should virtually eliminate the effects and sounds of echo in a national or international satellite circuit. Getting these relatively low cost devices (i.e. $500 to $1,000 per voice channel) systematically installed on a global basis must become another high priority goal of the satellite industry.

(c) <u>Enhanced Satellite Design and Performance</u>

Satellites have experienced enormous growth in capacity since 1965. In fact the INTELSAT I satellite is 170 times smaller in effective capacity than the INTELSAT 600 series. With digital compression techniques, the ratio will

FIGURE 18

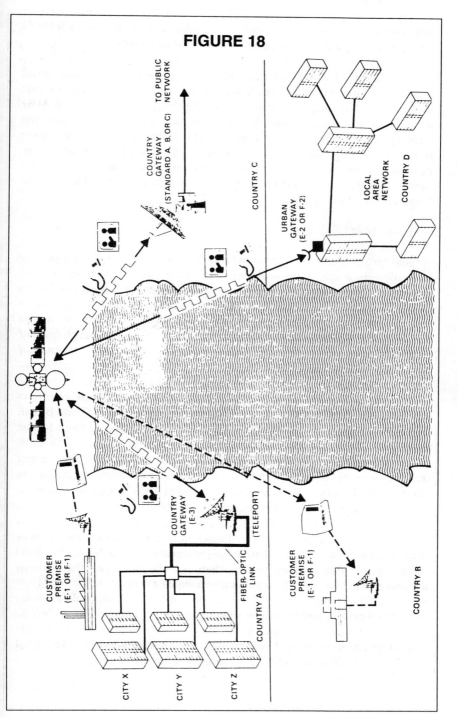

become 1 to 500 or 1 to 1,000. Satellite upgrades required for the age of ISDN and competition with fiber optic cables must go much further.

Satellites will need to provide downlinks that are 10 to 100 times more powerful in order to work to smaller earth station and provide margin against rain attenuation. Satellites will also need to last longer and offer a 36 MHz transponder at a price of about $50,000 sometime early in the Twenty-first Century. In short, the objective of the satellite industry should be to reclaim the technological leadership in the world of telecommunications by the start of the next millennium. To do so satellites will probably have to become 100 times more efficient. Figure 18 indicates that such objectives are not unreasonable based upon past experience.

(d) <u>Reasonable ISDN Standards</u>

A final objective, which will be key to making satellite voice service competitive and still very much in demand in the 1990's, is the adoption of fair and reasonable ISDN standards. These must allow all communication satellites to offer a full range of digital services at attractive rates. The most critical factor is to avoid the confusion between transmission problems associated with echo and delay. Echo can be solved with the right new equipment. The problem of delay has been proven in subjective tests involving tens of thousands of consumers to be largely a non-issue where single hop satellite voice communications is accompanied by high performance echo cancellers. Satellite-based ISDN voice service has, in fact, received astonishing ratings in consumer tests and demonstrations carried out by the Comsat Corporation AT&T and British Telecom. In short, all of the proposed ISD standards including those for the Broad band (or seventh layer) ISDN can be accommodated. The key element, however, is to retain CCITT recommendations (G. 114) with its finding that transmission times of up to 400 milli-seconds can be considered acceptable. As can be seen in Figures 19 and 20, delay is inherent in geosynchronous satellite operations and in fact plays a role in all satellite orbits.

Attempts to create a dual standard for satellites and fiber optic cable, such as some aspects of the SONET standard (Synchronous Optical Network) would unduly restrict media choice in transmissions and could divide the world of telecommunications into two separate and unequal worlds. Satellites and fiber optics would no longer be able to restore one another. International calls from the rest of the world, particularly developing countries which would largely (i.e., at least two-thirds) be incoming on satellite circuits would not be compatible with some proposed SONET delay-restricted terrestrial networks. The word "Integrated" is ISDN would quickly come to mean "Disintegrated".

FIGURE 19

TYPICAL VISTA APPLICATION

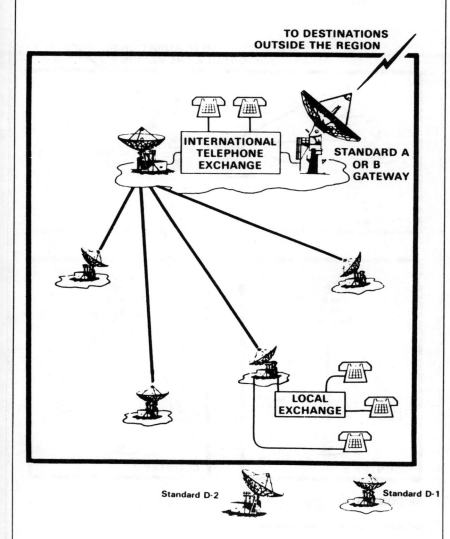

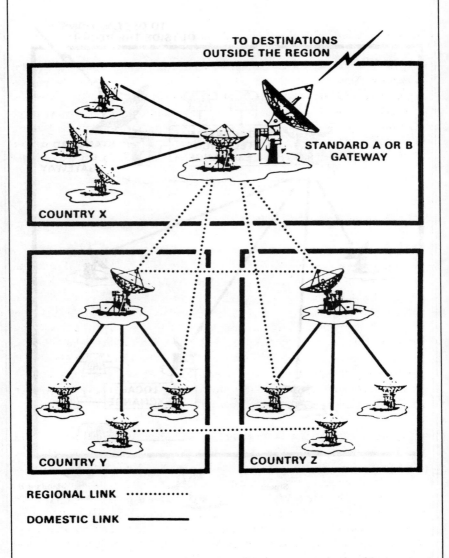

FIGURE 20

**DOMESTIC/REGIONAL VISTA NETWORK
WITH
STANDARD A OR B GATEWAY**

Fortunately, it seems not only possible but likely that the above four-point program will be successfully implemented by the satellite communications industry worldwide. ISDN, most likely will serve as a boon to the satellite industry by speeding the transition to high quality digital voice service and helping restore the prestige and subjective positive rating of satellites by consumers in the provision of telephone service at least internationally and on national VSAT dedicated business networks as well.

7.3 KEY ELEMENTS IN USING SATELLITE VOICE CIRCUITS

Anyone who does wish to use satellites for voice service should be an informed consumer. There is a good deal to know about the space segment, the ground segment and related terrestrial tails, and the optional voice-compatible services that can be obtained. On top of this there is the additional key issue of whether you want domestic satellite service in a particular country or whether you are talking about international overseas satellite service. Finally, there is the issue of whether you need public switched service, private network service, or both.

Let's sort through these aspects one by one:

7.3.1 National Versus International/Overseas Service

There are several basic ground rules to know. First, few national telecommunications agencies are now providing switched public voice service via national satellite systems. The exceptions are developing countries like Brazil, India and Indonesia which have their own systems or developing countries that are leasing capacity from the INTELSAT system for national service. Only in isolated areas like Northern Canada or Alaska does one tend to find public service being carried by satellite in developed countries. This situation may at a future date change for a variety of technical, operational, and financial reasons, but this is the general global pattern you find today. Most national satellite systems in developed countries are today either providing specialized networks for voice and data services or they are providing video services.

There are many reasons why this is true, but part of the reason is that many Post and Telecommunications Ministries as well as telecommunications giants like AT&T in the US., British Telecom in the UK, and KDD in Japan know that two-thirds of all overseas telecommunications is now provided by satellite. Accordingly, they wish to avoid "double-hopping" national public switched voice service into an international satellite circuit. Current ITU standards, in fact, do consider such service to be substandard, since the recommended maximum delay in a voice circuit has been set at 400 milliseconds. A tandem domestic and international satellite link is at least 600 milli-seconds.

The concern about satellite delay and double-hopping has also led to a new issue in the field of satellite communications which is often referred to as transborder satellite service. In this case domestic satellites with broader than national antenna coverage areas have moved to provide international services to adjacent countries. This, for instance, has led US satellite systems to seek to serve Canada, Mexico, Central America, and the Caribbean. Further the Canadian satellite system is also seeking to serve the US and the Caribbean. The Morelos satellite systems in Mexico is seeking to serve the US. In time, French satellites designed to serve French Overseas Territories in the Caribbean and off the shores of Canada and the US. may also seek to serve North America as a consequence of the satellite's antenna coverage.

The prime service area of interest for transborder satellites is, however, TV distribution. Although it is argued that the transborder satellite voice and data service is highly cost effective and not subject to potential double-hop delays as with domestic and international satellite interconnecting, virtually no public switched voice has migrated to transborder satellites. In fact, well under ten percent of revenues are from voice or data service in the case of transborder satellites and these are all private networks. The bulk of revenues from transborder satellite service is in the video area, and is likely to remain so.

7.3.2 Checklist for Evaluating Voice Satellite Networks

To ensure you make the best arrangements for voice circuits by satellite you should run through the following checklist:

(a) Analog Circuits

If you are going to use analog voice circuits you should have a good reason why. In the longer term they will cost you more, provide a lower quality of service and ultimately will be obsolete when ISDN is actually implemented on a wholesale basis in the 1990's. Make sure you have high quality echo cancellers installed on all lines you use, or if they are public switched lines monitor their performance and complain if substandard equipment or maintenance procedures are being used. Reasons why you might consider using analog equipment is to reduce front-end expenses, to have immediate "plug-in" service, or to achieve a "uniform" network when you have a combination of domestic and international service. The cost of a dedicated voice circuit on a satellite system including earth station costs start at about $1,000 per month and can range up to ten times that amount for some international service. The cost for some domestic analog circuits purchased in bulk in the US market can be obtained for well under $1000 per month.

(b) Digital Circuits

Although digital circuits may involve a higher initial investment in equipment, the longer term cost is less. This extra expense or investment covers the cost of digital echo canceller equipment and coding/decoding (CODEC) equipment, as well as 32 kilobit per second voice processing. Later in the 1990's it may be necessary or desirable to also equip for 16 kilobit per second low rate encoding. The net savings for digital service over analog in the long-term should be substantial particularly if one is buying the range of T-1 carriers (1.544 megabits per second) or the comparable European carrier of 2.048 megabit per second. The optimum configuration would clearly have to have a single dedicated digital network that operated from desk-top to terrestrial tail to earth station to satellite and back down through to the other work stations on the basis of a single modulation and coding process without being disrupted by digital to analog or digital to analog conversions. This is, in fact, what ISDN is all about.

(c) Ku Versus C-Band Satellite Service

In general, C-Band tends to provide higher quality service since it is less prone to rain fade. In most parts of the world, except Europe and Japan, C-Band can be cleared for earth station antennas in or near major urban centers. Ku-Band service can also be suitable if high-power satellites with high-gain antennas deliver an effective isotropic radiated power of 42 dBW or preferable above together with a suitable forward error correction scheme if digital modulation is being used. If it is indeed desired to provide service to rooftop antennas or connect to urban gateway antennas in major cities, in Japan or virtually all of Europe, then Ku-Band service is essentially the only option available. In fact, in Japan, the United Kingdom, and even the US, the trend is toward the use of Ka-Band (20/30 GHz) for customer-premise type service. In particular, Japan and the United Kingdom are very restricted in frequency clearance in all of their major urban areas in both C and Ku-Bands. This is because of extensive competing terrestrial microwave use.

(d) Access to the Space Segment

This is in many ways the key issue. There are, today, almost an infinite number of ways to get from the user office to the satellite. The most direct way is to have a desk-top data only antenna or a customer-premise earth station (CPE) in the parking lot or on top of the building with direct access to the satellite. The smallest such CPE antennas that handles voice service are about 2.4 meters in diameter and they cost in the range of $10,000 to $12,000. Some manufacturer's from around the world include NEC, Hughes Network Services, Scientific Atlanta, Comstream, Alcatel Telespace, Microtel, and about a dozen others.

The advantage of the customer-premise to customer-premise approach is total network control. Finding a radio interference-free site, installing the antenna and maintaining the customer premise earth station can be tricky if you do not own the building (e.g. a Manhattan high-rise may charge $25,000 to $50,000 in roof rent for a single dish). Maintaining a 24 hour-a-day repair and maintenance crew or even ready access to one is not cheap nor easy.

Furthermore, if you have in mind a combined national and international network involving locations in the United States, Europe, the Middle East, and Central and South America this could be difficult. You will probably seek to provide the service via the INTELSAT Business Service (IBS), which is essentially the only currently available operation. While the space segment will likely be relatively easy to obtain there may be big problems in using the customer premise approach. While the licensing and approval procedures by the FCC in the US takes some time, they are usually completed in several months. In the rest of the world the process may take years or may just not be allowed. Some countries are willing to license customers to operate premise earth stations for IBS as long as the government retains nominal ownership and control. Some will install customer premise equipment to users specifications, such as France Telecom. They will, however, insist on retaining full ownership, control, maintenance, and repair responsibilities, etc., and of course will charge appropriately. A diagram showing the various ways of obtaining international voice and data services is provided as Figure 21.

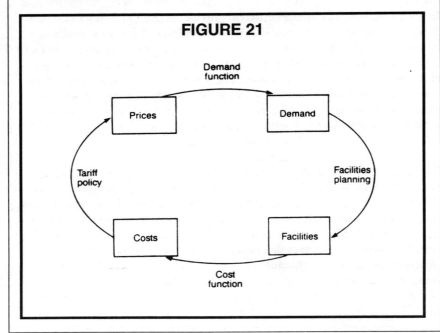

FIGURE 21

The long and short of this analysis is to advise flexibility in your planning. In the US the maximum number of opportunities exist to use customer premise equipment, to use terrestrial interconnection to an urban gateway station somewhere in the same town, or to use long distance tie-lines via microwave, coaxial cable, Wide Area Networks (WAN) or even fiber optic cables. The link into the satellite network may be at a location perhaps hundreds of miles away. In the US it is largely a matter of time, convenience, and cost. In other international locations it may be clear from the outset that only one option is realistic available because the others are too costly or simply are not available from a regulatory viewpoint.

It is seldom profitable to try to be the pioneer who changes the pattern of telecommunications regulation. If you do decide to try, here are some key words of advice: Try to get some local allies and keep another feasible option in reserve. Try to find a powerful firm, who is a non-competitor and is intent on achieving the same authorizations, then join forces. Finally, if you are planning to establish a large voice and other services network, bear in mind the importance of maintaining overall network reliability and control. You cannot effort a weak link in the chain. It is also good in this respect to recall Ockham' Razor: "If there are two or more solutions always pick the simplest".

(e) Pros and Cons of Establishing a Full-Time Voice Service Via Satellite Versus Other Options

Many corporations get mesmerized with all the buzz words: "Get in on the Telepower Revolution"! "Smart corporations control their future through Smart Networks", and so on. The idea of establishing a full-time national and international voice network that can be used in off-peak hours for data, electronic mail, facsimile, and even videoconference was one of the most oversold concepts of the 1980's. The only idea in telecommunications that is even more oversold is the idea that there are no risks associated with this idea, since you can always sell the leftover capacity to smaller corporation who can't afford to establish their own dedicated network.

Here are some basic truths to consider: (a) Leased public switched service particularly with enhanced service features can often be good buys, particularly if shrewdly purchased; (b) Piggybacking onto other dedicated networks even if some additional services must be met through the public-switched system can often by the best buy and can serve as a good interim strategy until you are indeed ready for you own dedicated system; (c) A "critical mass" of hard service requirements is often a good way of testing whether you are indeed ready for you own dedicated satellite voice system.

Here is a suggested "critical mass" checklist of deciding to proceed with a dedicated satellite system: (i) <u>Locations:</u> You should probably have at least four locations and preferably more to be served. (ii) <u>Traffic:</u> You should have enough traffic to support at least one T-1 (1.544 Megabits per second) carrier using 32 kilobits per second voice. This translates to 48 simultaneous voice channels or about 200 medium speed data channels at 9.6 bits per second or one or two compressed video channels using codec equipment from say NEC, Compression Labs, etc. (iii) <u>Security:</u> Security can be a crucial factor. When transmission security is critical then a dedicated network can make sense even with less traffic.

There are other circumstances, however, where even greater amounts of traffic still does not prove the benefits of establishing your own dedicated satellite network. The "critical mass" guideline outlined here is simply a good rule of thumb. Each case has its unique characteristics that must be considered. If you fall below 4 locations, cannot support a single T-1 carrier, and have no special security or quality requirements think twice.

(f) <u>Special Characteristics and Needs such as Security, Flexible Network Patterns, and Voice/Plus Services Like Remote Printing, Super High Quality Still Picture Distribution, etc.</u>

These are crucial factors. If you don't have some special characteristics to provide through your dedicated satellite voice network, then you may likely never wish to proceed further. But, on the other hand, if you do have such special requirements then a satellite network may well be for you. A digital satellite network can easily be augmented with encryption, special codecs or other special capabilities to meet special needs.

By using addressable earth stations you can vary the access at each site from full-time to a few minutes a day or even twice a week, if you like. You can also add or subtract earth stations at will with a minimum problems and perhaps little additional space segment charge. You could even operate several networks on the same capacity by effective time and frequency management. Spread spectrum technology, in particular, allows you to overlap one use over the top of another without interference, but this technique usually can provide data only and not voice service.

Remote printing or distribution of high-quality pictures for use in newspapers are applications of increasing use in conjunction with Bi-Phase Shift Keyed (BPSK) VSAT antennas that are also capable of voice communications as well. In summary, the establishment of a dedicated geographically extensive voice satellite system is an application that will continue even in an age of fiber optic cable systems. Nevertheless, it seems likely that most people who implement new satellite systems in the future will do so for voice-plus. It will be the "plus" features such as special encryption, network flexibility, and enhanced digital service features, that provide the basic rationale for establishing the new satellite networks.

7.4 RURAL AND REMOTE VOICE COMMUNICATIONS

The main focus of this chapter has been on urban-oriented satellite voice communications. Such networks, however, are essentially for developed countries like the US. The key issue concerning urban based business system tends to be the advantage of private business versus public networks. Satellite network for rural and remote services in both developing and developed countries, however, also have an important role. The truth is rural and remote voice services via satellite make sense. In many developing countries this is the only option technically available and is also the lowest cost solution.

Once a remote earth station, such as an INTELSAT VISTA D-1 terminal (4.5 meters and base cost of $60,000) is located in a rural and isolated region it can become the star or hub for a UHF terrestrial radio telephone network to extend the reach of network even further. Rural and remote satellite networks must keep in mind certain key features. These are: (i) antennas must be able to be powered by batteries or solar arrays because of unreliable or no electric power sources; (ii) earth stations and remote antennas must require little or no maintenance; (iii) antennas need to be as small as possible to aid in transportability, particularly to assist with stowing against tropical storms; etc.; (iv) antennas must be low cost. Today most rural and remote service such as VISTA (see Figure 22) is an analog service, but this service too has already begun to convert to digital as well. Tests on this issue, namely rural digital satellite services are now under way by INTELSAT and others.

7.5 CONCLUSION

If you want to establish a dedicated voice-plus satellite system or not, you will want to determine the relative cost and performance of all of the most likely options. One approach is to check with a variety of suppliers of service and ask them to cost out comparable systems for you. Another is to hire one of the many competent consultants who are available to perform this type of analysis. Consultants are usually readily at hand and can be found through the local, regional, or national offices of the IEEE, the Society of Satellite Professionals International, or through referrals by colleagues who have been through similar exercises, or even in a pinch the commercial section of your telephone book. My own rule of thumb is that a specialized satellite network should prove itself in by a 20 percent margin or forget about it. Of course, factors change and a reexamination every few years may prove worthwhile.

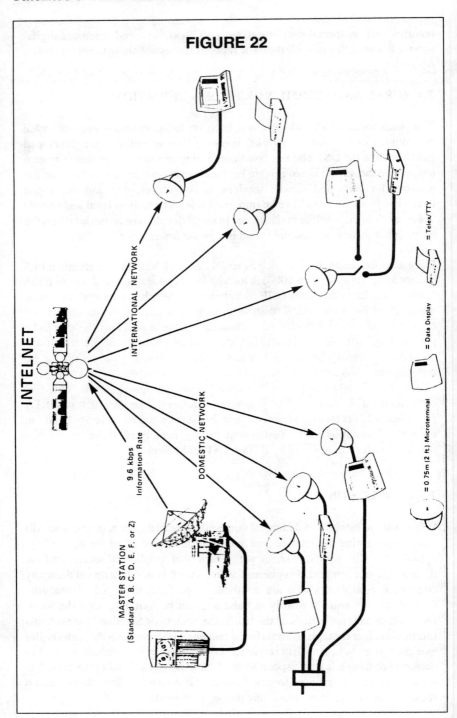

FIGURE 22

═══ CHAPTER 8 ═══

DATA COMMUNICATIONS AND SATELLITES

> "For large, variable destination, multi-mode data networks, satellite communications are very difficult to beat. In short, satellites are made for data networking."

Joseph N. Pelton
Future View

8.0 INTRODUCTION

Let's start with the basics. One might even say, let's review the obvious, but I think that would be going too far. It should be clear, for instance, that data communications services provided by analog transmission techniques, say Single Channel per Carrier (SCPC), is terribly inefficient. This should, in fact, be considered a "no, no" to be avoided if at all possible. If you have a data link to, say, one or two locations in South America or Africa and a satellite SCPC channel is the <u>only</u> option, then obviously you must accept what you can get. Nevertheless, digital satellite service is what you really want and need. Several options such as Time Division Multiple Access (TDMA), Intermediate Data Rate (IDR) carriers, IBS, etc., are now readily available to most locations internationally. These satellite services are ideal for data services of all types.

8.1 KEY STRATEGIES FOR SATELLITE DATA NETWORKS

In acquiring access to digital satellite circuits you probably want to consider the time factor. In particular you need to consider that the time and duration of access to be the critical factors for your decision. Data channels are frequently sold on a dedicated 24-hours-a-day basis, but can be time shared with others or sold hourly on a peak time or off-peak basis. Since user requirements for data transmission are typically not evenly spread over a daily cycle, this can translate into savings. There are, however, two key strategies that are often applied to solving what might be called the "peaks and valleys" problem:

- Combine Voice and Data Requirements: The most common solution within the business world is to combine daytime voice and other priority service requirements with evening and night-time data service requirements. In some cases, say, very large corporations, voice, data and even videoconference requirements are combined on T-1 carriers (1.544 megabits/second) or even larger 2.048 megabit/second carriers to give maximum opportunity for spreading of service requirements over the 24-hour period.

- Balance Real-time and Batch Process Requirements: Some data services such as telex, facsimile, and real-time data processing, say to support launch operations, are peak business hour requirements. There are, however, many data requirements such as electronic mail, digital voice mail, batch processing, file transfers and data dumps which are usually off-peak or at least non-priority requirements. Particularly, in an international data network where significant differences in time zones are involved, data communications requirements can be easily and cost-effectively accommodated to the overall 24-hour cycle.

In addition to these two basic strategies, it is also possible to buy additional capacity to cover seasonal peaks and to provide additional backup capacity if needed for certain critical periods while supporting a key-event or operation. A good MIS manager is always looking at a dedicated satellite data network as a valuable resource to be optimized so as to produce a maximum return on the investment. Brokers of satellite-time can aid this process.

8.2 DEDICATED SATELLITE NETWORKS vs. HYBRID SATELLITE/ TERRESTRIAL DATA SYSTEMS

Another basic issue to be faced from the outset is whether satellites, terrestrial transmission media or a combination of the two, make the most sense for your data requirements. There is today a myth, sometimes reinforced by advocates of fiber optic cables and wide area networks (WAN's), that satellite communications, because of their delay characteristics are unsuitable for data transmission. This is simply not true. Satellites meet all CCITT standards for all forms of telecommunications including data. The use of digital echo cancellers and of Satellite Delay Compensation Protocols has been successfully used by scores of sophisticated data communications users for both national and international data networks.

While some users will not put together tandem satellite hops, especially for combined national and international data networks, there are numerous users such as large banks, retailers, insurance companies, computer service companies, airlines, travel agencies, brokerage houses and many others who trust satellites for their data network transmission services.

Satellites provide high reliability and reduced costs often by eliminating terrestrial toll charges, by allowing instant on-demand access to the satellite capacity, and by creating almost infinite route flexibility that cannot be offered by terrestrial facilities. Satellites also allow the easy implementations of totally redundant end-to-end data links.

How does one decide whether to use satellites or terrestrial transmission techniques such as microwave, fiber optics or coaxial cable. The key decision elements for most customers are price, reliability, ready availability, and overall performance including instant access to all forms and types of services that are needed. If a telecommunications technology can't deliver in all four of these categories it is probably not for you.

Sometimes, however, it is a combination of space and terrestrial technologies which can deliver the best overall package. There are no magic formulas that guarantee the right answer.

There are, however, several guidelines that may help. One of the first things to consider is how big is the total geographic size of your network and how many nodes and pathways does it have in it.

If your network is located say within the LADA exchanges of greater Los Angeles or London and you have four nodes with 6 active pathways the choice is clearly for a terrestrial network. You might opt to plug into the Public Switch Telephone Network, perhaps by signing on for a full-time enhanced data channel. You might also build your own local area network (LAN) or plug into a privately operated wide area network (WAN) if it is available. In the US it is likely to be available, but in Europe or Asia it is not normally available under existing regulation.

On the other hand, if your network is in eight countries has thirty-three nodes and several hundred pathways, then satellite networking is again the obvious answer. You might find it is sensible to combine satellite and terrestrial access modes when you have two nodes in a single city or an existing terrestrial telecommunications system. This is only a matter of perfecting the details. The real issue is how to decide in cases that fall in-between. This issue of trying to decide when terrestrial systems should end and satellite systems should begin is never easy, often because the answers keep changing. There are no such things in the telecommunications industry as fixed long-term costs, since "future economic costs" of necessity are always in a state of flux. Figure 23 indicates the interaction of demand, facility planning, technology, cost and prices.

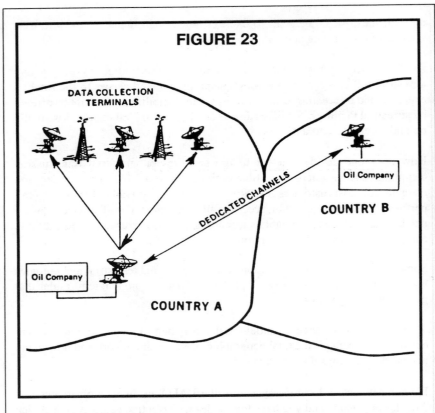

FIGURE 23

This rapid state of change and high degree of interaction among various factors is particularly characteristic of a high technology service industry such as satellites or fiber optic cables. Even so, satellite links are generally cost competitive for data connections over 500 miles.

8.3 TRADE OFFS BETWEEN SATELLITES AND FIBER OPTIC CABLES

Figure 24 shows the number of pathways in a network that can be created as you add nodes. It is assumed for convenience that in an "average" network about half of the possible pathways are actually used. In fact, use patterns are usually less. It is perhaps a useful guideline to observe that if the number of nodes increase above 5, if the number of "used" pathways increases above 10 and if the average pathway length exceeds 500 miles (or 800 kilometers) then a satellite data network is perhaps indicated; below these numbers a terrestrial network may be indicated. Only detailed system design and costing can of course tell you for sure.

FIGURE 24

Satellite vs Terrestrial
Evaluating Data Network Characteristics

Nodes	Possible Pathways	Average "used" Pathways
2	1	1
3	3	2
4	6	3
5	10	5
6	15	8
7	21	11
8	28	14
9	36	18
10	45	23
11	55	28
12	66	33
13	78	39
14	91	46
15	105	53
100	4,950	2,475
1,000	499,500	249,750

8.4 KEY SERVICE CONSIDERATIONS

Obviously, the number of nodes, the number of pathways actually used, and the average length of the pathways is critical to deciding how to proceed, but this is only the starting point. Your service requirements are vital. Key elements are: (a) transmit versus receive requirements (or if you like interactive data network versus data broadcast network); (b) transmission rates; (c) reliability and performance standards; and (d) earth station characteristics, assuming you opt for a space communications network.

8.4.1 One-way vs. Two-way vs. Interactive Networks

The basic type of data networking architecture is the most basic place to start. This seems logical since it represents the most fundamental consideration for a data user except perhaps for transmission speed.

<u>Broadcasting vs Networking:</u> The two things satellites do extremely well are broadcasting and networking. Strictly fixed point-to-point service is what cables and microwave do best. In satellite parlance broadcasting is point to multi-point service, while networking is multi-point to multi-point service. Of the two requirements, networking is the most demanding and from a satellite viewpoint it is the most expensive in terms of ground segment equipment.

One sophisticated installation can broadcast to hundreds or even thousands of earth station terminals. The spread spectrum type of earth station equipment using Coded Division Multiple Access (CDMA) can be used to send 19.2 kilobits of data to hundreds of terminals that are only 75 centimeters (2 feet) in diameter. This is called INTELNET in international applications as illustrated in Figure 25. These micro terminals sometimes called VSAT's for Very Small Aperture Antennas can be installed at remote sites for only $2,500 including microprocessor and printer. These terminals are so small they can be located on a desktop near a window with an exposure toward the equatorial plane or a view to the South. (In fact, if you are in the southern hemisphere, you would need a northern view to the equatorial plane).

The spread spectrum approach has many advantages such as high noise tolerance, ease of frequency clearance, low cost receiving terminals, and flexibility to add new locations at will. There are, however, two clear disadvantages. One is that hub earth stations that send out the data stream are highly sophisticated and expensive. The minimum cost of such a hub station is around $1 million. In short, you must have a lot of traffic to own and operate your own hub. This is the reason that so many satellite data network operators are opting for shared use hubs, whereby economies of scale can be more easily

FIGURE 25

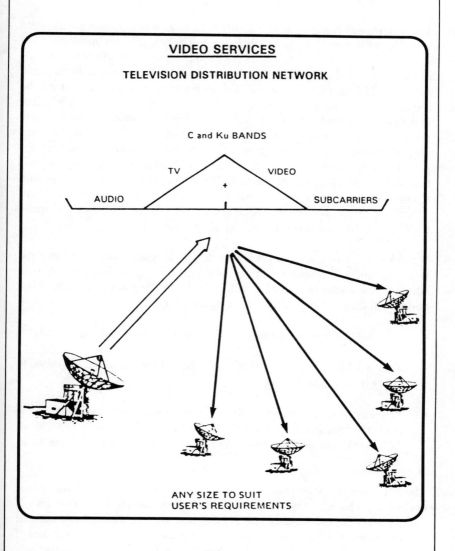

VIDEO SERVICES

TELEVISION DISTRIBUTION NETWORK

C and Ku BANDS

TV   VIDEO

+

AUDIO   SUBCARRIERS

ANY SIZE TO SUIT
USER'S REQUIREMENTS

achieved. As noted earlier, however, Mesh VSAT networks could overcome this problem and drive the cost of large satellite nets down from today's level. The ACTS experiments scheduled for 1992 should provided a good deal of new information on this subject.

The second issue is quality and reliability of service. When one designs a system with 75 centimeters (2 ft.) terminals at a cost of only $2,500 per installation, the savings in total ground system costs can be impressive, particularly if hundreds of locations may be involved. There is a down side associated with such cost savings. The bit error rate associated with such a ground segment can be typically expected to be in the range of 10-3 or possible 10-4. The BER is thus at least one thousand times worse than ISDN standards which are 10-7. There are techniques associated with re transmission and verification of all data than can, of course, improve performance dramatically. Likewise, special approaches to forward error correction (FEC) and higher rate encoding can also help. The point is look before you leap. Be certain you indeed wish to live with a point-to-multipoint broadcast system which utilizes very small micro terminals that may not give you the reliability and bit error performance you really want. It will be very expensive to upgrade at a later date.

For several reasons you may wish to design a system that gives you the flexibility to both transmit and network interactively within your system. Although the earth segment costs will undoubtedly be greater the system you design will likely offer you the following advantages:

• Your hub earth stations can be smaller, less sophisticated and cheaper.

• Your reliability and performance can be higher and in fact may be raised to ISDN standards if desired.

• You can communicate from all of the earth stations to all others in the network (typically these will be 2.4 meter (7 feet) antennas that cost about $12k-$15k and use Bi-Phase Shift Keyed (BPSK) transmission techniques).

The real issue is: Will your network evolve to new services? Even if it is exclusively a broadcast network today, will it continue to be one tomorrow? If your network contains twenty or less earth stations in it, then consideration of a full up interactive network that uses 2.4 meter antennas makes a great deal of sense. Above that figure then, a strictly broadcast network or possibly a partially interactive network as illustrated in Figure 26 makes sense. As a final footnote on this subject, it should be noted that an interactive, higher performance spread spectrum antenna is also available on the market starting with a 1.2 meter

FIGURE 26

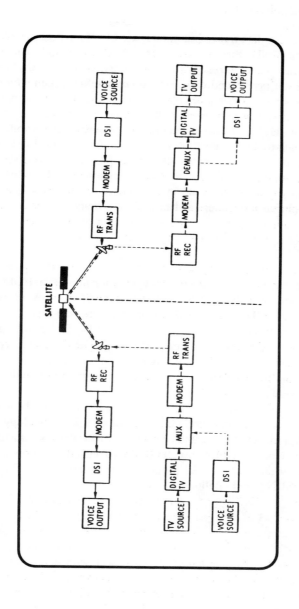

antenna which can receive at 19.2 kilobit/second and also transmit at 2.4 kilobits per second. This type of VSAT costs on the order of $8K.

8.4.2 Transmission Rates

Requirements for data transmission rates are of course determined by actual service needs. Today these usually start with what used to be called medium speed computer rates of 9.6 kilobits per second. This is followed by higher speed 56 or 64 kilobit per second data rates to cover either digital voice or rapid exchange between medium speed processors. Above this level comes the truly high speed data exchange between high speed processors either at 1.5 megabits per second in the US (a T-1 carrier) and 2.0 megabit per second in Europe. Finally demand for even higher rates are beginning to materialize especially in the US Government and research institutions plus a few businesses are now using DS-3 data links which operate at 45 megabits/second. Definition of the new broad band ISDN standard at 155 megabit/second and 620 megabits/second will likely open up applications at these truly wide band rates in the late 1990's.

8.5 CONCLUSION

Satellite communications with delay compensation protocols (HDLC) are now able to provide virtually error-free data networks of all types. It can support satellite based LANs for corporate customers. One-way broadcast, two-way links or even multi-point to multi-point data networks of great size, complexity, and variable rate can certainly all be provided. Technological innovations are serving to create very small and inexpensive satellite data networks. Digital modulation and encoding techniques also allow for improved quality, security, and system availability. Although cost trade-off analysis between satellites and fiber optic cable systems clearly are changing with great speed due to techno-logical innovations, there is no indication that satellite technology is lagging behind. In particular, in the area of data networks, there seems likely to be continued growth of satellite data networks. This will be in the context of hybrid satellite/terrestrial data networks, private data networks on VSAT systems and even some ISDN based satellite systems as well.

CHAPTER 9

SATELLITES AND VIDEO SERVICES

> **"Satellite television is the greatest invention to date for the super tribalization of man."**

Desmond Morris
Anthropologist

9.0 INTRODUCTION

When Marshall McLuhan coined the phrase the "Global Electronic Village," he considered television to be the revolutionary technology that would make it all happen. Nearly twenty years later the futurist John Naisbitt in the best seller <u>Megatrends</u> attempted to set the record straight. The radical new invention that allows the creation of the "Global Electronic Village" in Naisbitt's view as not television, but the satellite. Only the satellite allows instantaneous global audience to become possible in developing as well as developed countries. Desmond Morris the pop social anthropologist has said the same thing in a totally different way. He claims that the satellite is the single most potent force for "Super Tribalization" yet invented by man.

Who knows. Perhaps Naisbitt and Morris were both right. Television and the satellite may together reshape the world and our perception of it. Consider this brief history of global satellite television.

<u>Moon Landing, July 1969:</u> The first global television show with simultaneous television distribution in the Atlantic, Pacific and Indian Ocean regions was the manned moon landing. A record 500 million viewers watched worldwide "Live via Satellite."

<u>Summer Olympics, Montreal, Canada; Summer of 1976:</u> The most extensive coverage of any television event in history was the Montreal Summer Olympics when 1 billion people tuned in via satellite. This record did not last. The Seoul Olympics reached over 40% of the world's population with over 2 billion viewers. Over 30 transportable and permanent earth stations uplinked close to 100 different TV channels to bring the Korean Olympics to the world.

<u>Full-Time International Satellite Television:</u> Since 1984 there has been a worldwide explosion in international satellite distribution. The number of full international television channels on INTELSAT has increased from zero to over fifty. Regional channels on EUTELSAT satellite and ASTRA are also expanding rapidly to provide dozens of channels in Europe. PANAMSAT, ORION and ASIASAT are beginning to offer expanded video options. The next two steps will be (a) global full-time TV as opposed to regional coverage TV distribution. (In fact, CNN has already accomplished a near global TV channel with nearly 100 countries now reached by satellite TV distribution in 3-ocean regions), and (b) international DBS TV programming based upon the now almost standard formula of movies, sports and news, will become available before the 21st century.

<u>Global Satellite Tele-education:</u> Global satellite tele-education and tele-health represents one of the latest key trends in Global Satellite Video Services. China has set up a National TV University in only 2 years. It now has 5,000 receive only terminals and over 1 million students. The INSAT project in India is even larger and has been in existence for over 5 years. The potential of extending national networks into regional or global systems is enormous. This trend is now under way with the Commonwealth of Learning, TV Ontario, and the Knowledge Network of the West projects out of Canada. In the US. there a large number of tele-education initiatives which include the International University Consortium, the International Space University, the National University Tele-conference Network (NUTN), SCOLA, the University of the World, the Foundation for International Tele-Education (FIT), and the Global (Pacific) University. The National Technological University has recently indicated it will also consider international course offerings.

Dozens of projects in fact exist in other countries around the world such as the Telekolleg of Germany, the Open University and the Educational Broadcast Services Trust in the United Kingdom and so on around the world. This is certainly not just a developed world phenomena with exciting projects such as the University of the West Indies and the University of the South Pacific showing an excellent record of satellite education for a period that is now two decades in length.

Finally there is the issue of non-formal education via satellite distribution. Besides clearly educational satellite distribution, there are news, sports, entertainment channels which still impart a great deal of knowledge and these channels which reach a much larger audience may be making a much bigger impact. Global video satellite distribution will if fact become a multi billion industry in the 21st century and its impact on world education must also be considered.

9.1 THE MERITS OF SATELLITE TELEVISION

What the above examples of dynamic growth in satellite television help to substantiate is this simple fact: Television and satellites are made for each other. The evidence is clear:

- Television is the mainstay source of revenue for satellite service. (It is often of more of all service revenues for US satellite systems).

- Television is the top growth service for international satellite systems.

- Television is the service that the public associates with satellites in terms of back yard dishes, of "live via satellite" broadcasts, etc.

- Television by satellite is cheap, effective, high quality and not adversely affected by satellite delays. Further, it becomes more and more efficient as viewers and coverage grows, particularly for rural and remote areas.

- What the new age of HDTV arrives, satellites will be excellently positioned to provide this new offering in the US. and around the world.

9.2 SATELLITE TELEVISION - KEY ELEMENTS FOR NOW AND THE FUTURE

In light of the enormous potential of satellite television, there is a need for clear strategy for this market both now and in the future. This means examining strategies that involve conventional analog television; new television standards; digital television; high definition television and longer term prospects such as 3-D television and holovision.

9.2.1 Analog Television

Today perhaps 85% percent of all terrestrial and satellite distributed television is sent as analog signals. This is not surprising since analog TV is reliable, proven, inexpensive and is widely available in most parts of the world. Digital capability, on the other hand, has only just begun to become established. In short, while digital satellite television is clearly the service of the future, analog is today's technology.

To establish a cable TV satellite a cable TV satellite distribution service, a full-time video news channel, or some other commercial TV distribution network today, analog may be still the way to go. This is because it "fits in" with the existing earth station networks, the available satellite capacity and the home TV receivers and local distribution systems. There are several satellites for instance which are "hot TV birds." These have 8, 10, or even 12 full-time TV channels on them. Thus, a single antenna can be used to receive all these TV channels. The incremental expense for receiving an additional analog TV channel is very small.

Digital TV is a complication. It is thus easier to create a new analog TV channel for a national, regional or international service. A shift to digital technology would likely only make sense as part of a major step function change with 6 or more TV services switching over all at the same time. A new digital TV coalition, or as in the case of DirecTV, a very large and well financed company, would in effect jump off the digital cliff together.

Assuming analog TV is here for a while, there are several issues that are important to consider:

9.2.1.1 Signal Quality (Signal-to-Noise S/N)

Broadcasters worry a lot about quality. The quality of program content is supreme but high picture-quality is still important. It is the combination of the two that the Broadcaster sells. Much more money is spent on new program production, however, than on transmitting the program to the general public. Thus, there is logic in saying: "Let's keep our distribution network and quality high since it is less than 10 percent of our total costs". In short, why worry about saving pennies on broadcast transmission quality when you are spending big bucks on program production.

There are only a few key elements of an analog TV transmissions that affect picture quality. These are the power, the frequency and the transmit gain from the satellites, the Gain-to-Noise rations (G/T) of the receiving antenna and the terrestrial system for getting the TV signal from the earth station antenna to the TV transmitter or cable TV headend for redistribution. The idea is to use all of these dimensions with sufficient margin to create a "good" picture for the home viewers. Bearing these functions in mind, let's consider the key components.

9.2.3 Frequency Bands

Carriers that vary from 17.5 MHz to 72 MHz are used for satellite TV. INTELSAT, which used 17.5 MHz carriers for international TV found them too low in quality and went to 20 MHz carriers (two per 41 MHz transponder) when the INTELSAT V service became available. It is generally accepted that carriers in the 20 to 36 MHz range are optional. 72 MHz carriers can provide excellent quality, but are often considered wasteful of the spectrum.

9.2.4 Power

Satellite transmit power which is determined by available on-board power and antenna gain is crucial to good quality. The minimal acceptable power for most satellite television transmissions in the C-Band (6/4 GHz) is around 33 or 34 dBW for national TV distribution purposes although bandwidth can be traded off against lower power. Today most systems, particularly in the Ku-Band, are in the 44 to 48 dBW range. The INTELSAT system which is global in coverage has typically used lower levels, but they too are now upgrading with 6 dB higher power and offering higher transmit power of 36 dBW in the C-Band zone and hemisphere beams and up to 44 dBW in the Ku-Band spot beam in the new INTELSAT VII satellites to be launched in 1993. Figures 27 and 28 give and indication of EIRP values to be found in C-Band and Ku-Band satellites.

Because of rain fade and the need for increased power to protect against it, the Ku-Band transmit power levels should be 6 to 8 dBW higher than in C-Band. Most "received" TV signals are, today, in the "very good" range of 52 to 56 dB Signal-to-Noise (S/N). In some cases involving TV carriers for remote and rural TV distribution, however, Signal-to-Noise rates as low as 45 dB have been used. A general representation of a TV distribution network is given in Figure 25.

9.2.5 Receiving Antennas

Receiving antennas for TV distribution today are sized anywhere from 3 to 11 meters (20 dB/K to 35.7 dB/K). Ku-Band antennas tend to be in the 3 to 7 meter range while the C-Band antennas are typically 4 to 11 meters. The larger TV antennas in the 9 to 11 meter range are usually for the reception or in some cases reception and transmission of international TV to the lower powered INTELSAT satellites. Most antennas are terminals used only for reception and are thus quite inexpensive, i.e., from $600 to $1500 while the transmit and receive antennas are much more expensive (i.e. from $100,000 to $1,000,000). Some terminals are in fact designed to be quickly augmented using mobile equipment to become a temporary transmit facility.

There is one element of the TV antenna issue that will become a key point for operators of TV satellite networks. This quite simply is the issue of inclined orbit satellites. These satellites can produce good power levels and high quality (as measured in Signal-to-Noise tests). The key is calculating the financial tradeoff concerning the use of such satellites or not. Inclined orbit satellites provide you with lower space segment costs, but unless you already have a

FIGURE 27

TYPICAL EIRP MAP C-Band
This example is:
Nahuel Beam A

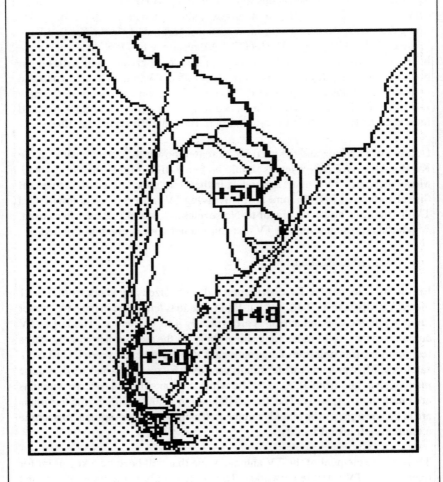

FIGURE 28

TYPICAL EIRP MAP Ku-Band
This example is:
TDF-1

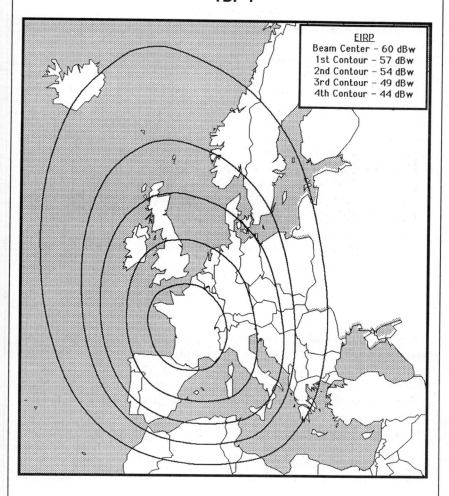

EIRP
Beam Center – 60 dBw
1st Contour – 57 dBw
2nd Contour – 54 dBw
3rd Contour – 49 dBw
4th Contour – 44 dBw

sophisticated earth station network with built-in tracking, you can have substantial earth station retrofit costs. A rather simplistic formula can give you a rough indication of what the net gain or loss may be.

$$5(CSR-CSI)=(RCI)NT1 + G$$

This formula is based upon a 5-year period for retrofit cost pay back. You can vary this, however, by increasing or decreasing the first number in the formula, i.e. inserting 7 for 5 years or perhaps 4 in place of 5 years. This formula can be explained as follows. The annual cost of the regular satellite time now in use "CSR" less the annual cost of the inclined orbit satellite "CSI" gives the annual cost reduction for using the inclined orbit satellite. On the other side of the formula is the cost of retrofitting. The retrofit cost of the antennas of type one "RCI" is multiplied by the number of antennas of that type "NT1". This process continues for all the various types of antennas in the system. If for ease of analysis we assume that there are 50 antennas, all of the same type and the retrofit cost is $20K, then the calculation of the net gain G is very straight forward. If the annual space charges are $1.5M for the regular service and 0.75M for the inclined orbit capacity then the solution for G or Net Gain is as follows:

Sample Solution of Cost Benefit for Inclined Orbit Retrofit

$$5(\$1.5M-0,75M) = (\$20K)50 + G$$
$$\$3.75M-\$1M = G$$
$$\$2.75M = G$$

This is a simplistic analysis. The present value of money, for instance, has not been considered, but obviously a $2.75M "rough-cut estimate" of the likely payoff in 5 years suggests the retrofit is probably well worth it. In short it certainly seems worth it to spend $1 million in retrofits in order to save $3.75 million over the next five years and to realize a net gain of some $2.75 million. This of course ignores net present worth but even after this is taken into account there is still a large and clear savings. Most cases won't be equally clear cut.

9.2.6 New Television Standards

The number of TV standards in use for analog television today is somewhere between incredible and appalling. First there are the basic three TV standards of (a) The National Television Standards Committee (NTSC) developed by RCA/NBC and used primarily in North America, Japan and South and Central America; (b) the French standard originally developed by Thomson CSF known as SECAM that is used in Africa, other parts of Europe, the USSR; and (c) PAL the German standard developed by Siemens and used extensively within the Commonwealth countries (UK, India, Pakistan, Sri Lanka and other locations in Europe).

In addition to the big three TV standards there are various forms of the Multiple Amplitude Component (MAC) standards which are used with satellite transmission. There are many forms of MAC which relate to different audio channels B-MAC; C-MAC; D-MAC and D-2-MAC. The MAC standard as developed by the Independent Broadcast Authority (IBA) of the United Kingdom and can be adapted to work to NTSC; PAL and SECAM TV sets which is one of its chief advantages. Even so, the narrow audio options on MAC transmissions leads to a new set of standards issues.

TV standards and satellite communications are the biggest problem when international TV exchanges are involved. A three country interactive TV hookup just involving the US, the Federal Republic of Germany and France means TV standards conversions as follows: NTSC to PAL; NTSC to SECAM; PAL to NTSC; PAL to SECAM; SECAM to NTSC, and SECAM to PAL. These sixfold conversions do not work equally well. The NTSC standards at 525 lines and 60 cycles per second is a more grainy picture at the outset. When NTSC is converted into PAL or SECAM which is a 625 line and 50 cycle per second picture, the result is definitely inferior. The other conversions to NTSC from SECAM or PAL fortunately produces better results.

Planning for international TV broadcasts must therefore consider what standards conversion is necessary, who is equipped to do them and which is the best place in terms of cost, expertise and experience. Some groups specialize in satellite TV standards conversion. VISNEWS in the suburbs of London for instance is perhaps the most sophisticated house available to do any standards conversion required and is also equipped for computerized enhancement of the image once converted. Other entities such as the German Telecom Organization specialize is routine conversion of PAL format TV coming from Asia and relaying it in converted standard to North American or Europe. In any event standards conversion is a key part of the process of putting together international satellite TV programs.

9.2.7 Digital Television

The rate at which satellite TV switches from analog to digital will be driven by several factors such as: the speed with which new digital standard are set; how quickly digital TV is offered at attractive rates on national and international fiber optic cable; and how seriously broadcasters take up the opportunities presented by Broadband ISDN or decide to go with a dedicated private system.

The early start of DirecTV, USSB and PrimeStar in the USA has started to fuel a move toward digital television. The satellite industry will need to be nimble to not be left behind. Satellite operators may well wish to work with broadcasters and cable TV planners to move a half dozen or more channels from analog to digital all at the same time.

The world of television, at least analog television, could be said "to belong" to satellites today at the national, regional and global level. There is some danger, if the satellite industry does not move more swiftly to digital television via satellite in the late 1990's. If this does not happen then a trend line might go with fiber optic cables.

Furthermore, it could serve as a "bell weather" for other digital services as well. In light of the future critical nature of the digital transmission of TV, satellite operators without direct investment in ground segment may wish to consider various incentives to help with antennas retrofits to keep broadcasters in the fold. As Figure 29 shows the addition of digital TV capabilities to existing earth station antennas is a reasonably straight forward step to take.

9.2.8 High Definition Television and Beyond

There are two main stimuli for high definition TV. These are the low quality of the NTSC standard with its poor 525 line resolution and the commercial interest of TV set manufacturers who would like to create a higher growth market beyond that of just replacing old color TV receivers. Given this motivation it is not surprising that Japan and the US are at the forefront of the ranks of those seeking a new HDTV standard. The initial efforts were to develop a new 1125 line system known as MUSE with a 9 x 16 aspect ratio. These early efforts were largely undertaken by NHK and SONY of Japan.

Europeans with current TV standards based upon 625 line systems are more interested in upgrading their existing receivers with line doubling techniques to get 1,250 line resolution. In the US some wish to go on "advanced television" with today's 3 x 4 aspect ratio and a less ambitious new resolution. The problem is that technically there are no end of possible solutions, and politically US, Japanese and European television manufacturers and programmers have little incentive to agree.

If one visits MIT's Media Lab in Cambridge, Massachusetts, they will find a 2,000-line by 2,000-line monitor built on an experimental basis by SONY which is on trial there. Its resolution is breathtakingly sharp — almost beyond reality. At 60 frames a second, one would require enormously high throughout of perhaps 1 or 2 Gigabits per second to transmit pictures to such a super HDTV

FIGURE 29

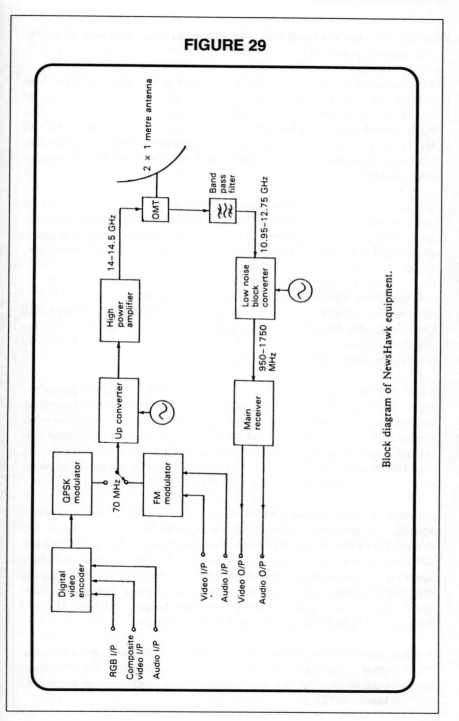

Block diagram of NewsHawk equipment.

screen. Of course, digital compressions techniques would pull this down to 200-500 megabits per second.

This Super HDTV demo project serves to highlight another aspect of high definition television. The tremendously high rate of digital throughput of HDTV should be very good business for communication satellite systems. The future growth of satellite digital TV will likely be largely built on the basis of HDTV. It is, in fact, the only truly wide band service on the horizon other than computer based imaging where interactive CAD/CAM and super computers interconnect.

The keys to success for satellite-based HDTV will be to cover all the bases in terms of covering all possible HDTV standards, work out attractive tariffs, developing transmission plans and carrying out tests and demonstrations to attract HDTV service to satellites early. Further, one must stay alert to new trends.

One possibility is the conversion of motion picture chains to large screen HDTV distribution centers perhaps offering discount rates. The satellite industry which could make this happen quickly could be in on the ground floor. Another possible implementation concept of HDTV would be as an add-on feature for a Direct Broadcast Satellite (DBS) System.

Most conventional satellite systems are not equipped in terms of power and frequency for the high speed digital throughput needed to send many hundreds of megabits per second. A specially designed DBS satellite working together with a digitally compressed signal could, however, provide affordable offer home based HDTV. This would be one of the few truly new DBS services that could be offered. Further it could offer truly total and instantaneous coverage in a way that a terrestrial fiber optic system could not duplicate.

Another concern is that of HDTV service actually "wasting" radio frequency spectrum. The first HDTV standard, the Japanese Muse system was in fact designed to be distributed via two very wide band transponders on the Japanese BS-1 direct broadcast satellite. The assumed easy availability of wide band spectrum certainly served to create an HDTV standard that was not frequency efficient. The later HDTV systems, many of which were designed for cable television or conventional over the air television channels of either 6 Mhz or 9 Mhz are clearly much more capable of saving frequency.

Satellite demonstration are thus badly needed to show that the issue of spectrum is not related to a particular transmission mode and that satellites are indeed quite compatible with spectrum efficient versions of HDTV and in fact are compatible with all known HDTV formats.

Where the trend for higher resolution video will stop is hard to say. This much seems likely, higher quality video is not likely to be power or bandwidth constrained. Satellites and fiber optic systems will probably reach transmission speeds of 100 Gigabits per second between 2010 and 2020. These rates should be able to handle super high resolution video, 3D TV or even some form of holovision scaled to human viewing needs.

Certainly, the next frontier for wide band communications beyond HDTV is likely to be some form of three dimension television. There is research now under way to film or shoot video images from say 7 different fields of view and project them to create a "solid looking" image. This is sometimes called multiple-raster-imaging.

Even though these could probably increase throughput requirements by a factor of 10, this is still a very interesting area of inquiry, particularly as transmission costs seem likely to drop by a factor of at least 5 times each decade.

Probably the ultimate in throughput for the Twenty-first Century could be Holovision. Even scaled to human viewing range of 40 degrees horizontally and 40 degrees vertically, a Holovision transmission system would probably require all of 100 Gigabit per second - a speed not yet available on planet Earth, but very likely to be available early in the next century. Such advanced systems may likely be developed for entertainment purposes, but the health, medical and educational benefits may also prove to be enormous. Certainly they can and likely will help sustain a healthy satellite communications industry into the twenty-first Century.

9.3 CONCLUSION

The satellite TV distribution system is today at the very core of the overall satellite communications industry. It is not only a key source of revenue but it is the very "foundation of wide band service". The ability of the satellite to offer consistently reliable telecommunications service with an effective economy of scale derives first and foremost from television, especially in the various national satellite communications systems.

It is for this reason that the current transition period is so crucial. If in the change over from analog to digital, satellite TV should happen to develop a sneeze, the satellite industry will catch a bad cold. The same sensitivity applies to new ISDN reliability and service standards. Clearly, the shift from today's TV standards to HDTV is strategically important to the satellite industry.

If satellites should for technical, operational or cost-related reasons somehow slip and give up a substantial share of the market to fiber optic cable, it could be a major blow to the satellite industry. Clearly the satellite industry should stay closely linked to the broadcasting and cable TV industry. It should cover all the bases and be a quick study on all the issues: ISDN quality standards; standards conversions; 1125 and 1250 line HDTV; multiple rastered TV; and analog and digital services are just some of the critical areas.

To remain competitive satellite systems will likely need to actively market promotional rates and incentives for wholesale conversions from analog and digital with subsidies for earth segment conversion equipment.

Finally, inclined orbit satellites could constitute another dimension of this critical transition period although of a much less pivotal nature. Somehow if satellite planners can provide good coverage, higher power, digital reliability, and tariff flexibility, satellite video will still be on top in the twenty-first century.

It is difficult to spell out in detail every aspect of a satellite network strategy for the late 1990's. The first step is to recognize that such a strategy is of critical importance, and to start working on it now.

CHAPTER 10

KEY NEW TECHNOLOGIES

INTERSATELLITE LINKS AND HIGH ALTITUDE LONG ENDURANCE PLATFORMS

> **"The most constant fact of life in the satellite telecommunications market is sustained growth and development."**

George Lissandrello
Columbia Communications Corporation

10.0 INTRODUCTION

The field of satellite communications since its beginning over 30 years ago has been in constant flux and development. This industry has developed dozens of new technologies and systems that has allowed performance to increase more than a thousand-fold. The rapid parallel development of fiber optic communications has created strong incentive for the satellite communications field to keep forging ahead at a rapid pace.

There are literally dozens of prospective technologies that are now on the horizon, but two technologies now being developed and implemented could make truly significant changes in the next five years. The first of these is the inter-satellite link which can be used in a wide range of ways for many different satellite systems. The Orion geosynchronous satellite system and the Mototola and Teledesic low earth orbit satellite systems are all committed to these ISL links and many more systems will follow suit.

The second fascinated technology is not even a satellite technology at all. It is a platform that can provide wide area coverage telecommunications service from so-called "proto-space". These remotely piloted vehicles or unattended autonomous vehicles are deployed above commercial aviation space allows a cost-effective way for the provision of wide area cellular, PCS, or video services. These platforms which cost only 1% to 2% of a satellite system can be operated over an area of 181,000 sq. kilometers as an independent unit or it can be creatively combined with satellite systems to interconnect with other networks or relieve satellite system congestion. These two technologies should both make a major impact even before the close of the decade.

10.1 INTERSATELLITE LINKS: THE NEXT KEY FRONTIER

The intersatellite link (ISL) is a means of interconnecting by electromagnetic means two or more artificial satellites in earth orbit. In the future, intersatellite links could be used to interconnect artificial satellites circling the moon, other planets or even the sun. These links can vary in telecommunications or information carrying capacity, and they can be used to connect satellites that are close together or far apart. The purpose for which an intersatellite link is used can also vary but frequently it for one of the following purposes: (a) The relay of scientific or earth monitoring data from low earth orbiting satellites to an earth based data processing center. This is often accomplished by relay of the information via an intersatellite link to a geosynchronous satellite in a stationary and much higher orbit; (b) A second application is to interconnect two satellites which are closely positioned together in order to redistribute traffic and maintain traffic interconnectivity within a particular region; (c) A third application is to interconnect directly two satellites which are serving different regions of the world and thus avoid a "double hop" link that requires a signal to be sent to and from satellites two different times to make the required linkages; (d) Finally a large constellation of satellites in low earth orbit can be interconnected by intersatellite links to create a global mobile satellite system of high efficiency. This also allows a high degree of radio spectrum efficiency due to multiple frequency reuse techniques that call be derived from low-orbit systems.

Intersatellite links tend to operate at the very highest radio frequencies in the Super High Frequency (SHF) band (i.e. the millimeter wave band) or at optical or lightwave frequencies even though the initial systems have used frequencies as low as the S-Band. The use of the higher frequency bands makes considerable sense in that there is a broad band-width available and the problem of precipitation attenuation goes away above the earth's atmosphere.

For many years the concept of Intersatellite Links seemed to a theoretical possibility for future satellite applications. ISL operation began with the Tracking and Data Relay Satellite System launched by the US. National Aeronautical and Space Administration (NASA) and the Loutch system launched by Russia starting in the second half of the 1980's. From this limited beginning, the practical applications of ISL are beginning to grow. Today a large number of satellite systems have indicated plans to ISLs for governmental, military, and commercial purposes.

10.2 THE FOUR BASIC TYPES OF ISLs.

The basic idea of intersatellite links is to achieve interconnectivity between satellites. This can be for convenience, cost savings, reduction of transmission delay, or meeting of a basic service needs in the most efficient or in some cases the only possible way. The four basic types of ISLs are described here to help show they operate and to explain how they are different from each other.

10.2.1 ISLs for Co-Located Satellites

This is the simplest form of ISL. In this case, the satellites to be connected are usually less than 80 kilometers or 50 miles apart. Typically the satellites would be in geosynchronous orbit some 35,870 kilometers (22,238 miles) above the earth's surface. This technique can be used to add capacity where an existing satellite has exceeded its telecommunications carrying capacity and certain earth station-to-earth station direct linkages need to be maintained. Feasibility studies have been carried out to show that these relatively short connections could actually be provided by fiber optic cables as well as the more conventional use of microwave radio frequency (RF) interconnections. An extension of this type is the intraregional ISL where satellites are in geosynchronous orbit, but separated by much greater differences such as 5°, 10°, or even 15°. These satellites, which are still operating within the same region are interconnected to maintain desired earth station-to-earth station direct linkages without resorting to double hop re-transmissions. These ISL's unlike the case of virtually co-located satellites require more sophistication because of the much greater distances involved with a 5° separation representing a link of 3850 kilometers (or 2425 miles), a 10° separation representing 7700 kilometers (or 4850 miles) and a 15° separation representing 11,550 kilometers (or 7,275 miles). With all ISL's, the greater the distance of separation there is a need for more transmit power, more sophisticated and higher gain antennas for transmission and reception, and more accurate tracking and pointing mechanisms.

10.2.2. Interregional Inter Satellite Links.

This type of intersatellite involves connection across enormous distances. From a technical point of view it is very difficult to accomplish for a variety of reasons. These reasons include the need for the following capabilities:

a) High speed processors capable of instantaneously calculating the location of the satellites to which ISL communications is to be established.

b) ISL antennas capable aiming an RF beam of 1° width or less or better still an optical laser beam directly toward the targeted satellites at remote locations.

c) Sufficient on-board power to support the communications satellite's mission for earth-to-space and space-to-earth communications as well as to support the ISL's power needs too.

d) A communications processing and switching capability with sufficient speed and power so as to allow an effective use of this capability in the routing of calls and data links to the intended location.

There are limits to the connectivity in inter-regional ISL's due to the earth's obstruction of the pathway. The so-called zone of exclusion requires at least three satellites equipped with ISL's to achieve total world-wide interconnectivity. In short separations of about 100° is close to the practical limits of the interconnectivity that can be achieved in an inter-regional ISL communications satellites.

In general terms the ISL technology will tend to move from the RF frequencies to optical laser links in the late 1990's. This is because the much greater bandwidths that lightwaves can provide will be needed to handle a much greater volume of traffic. Further, the reliability of laser diodes and other components of optical ISL's is greatly improving while mass requirements are decreasing. Finally the experience with experimental optical ISL systems such as the Japanese Experimental Test Satellite VI, the European Space Agency's Semi-Conductor Laser Experiment (SILEX), the European Data Relay Satellite 1, and the NASA Laser Communications Transceiver will demonstrate all suggest that this technology will mature before the year 2000. Military tests will examine how ISL technology can be used as strategy to avoid or minimize the effects of jamming.

10.2.3 Low-Orbit to Geosynchronous Orbit Relay

This type of ISL relay is most extensively utilized at this time. It has been used by the United States to relay data from the Hubble Telescope, the Shuttle manned spacecraft missions, and a growing number of low orbit satellite networks. As satellite networks with large data relay systems expand such as the Earth Observation System, the upgraded Search and Rescue Satellite system, develop the need for this type of ISL capability will expand along with the throughput requirements. The Russian Loutch system has demonstrated similar capabilities and European and Japanese capabilities will be demonstrated in the late 1990's.

What is unique about this application of ISL capability is that there is no easy substitute for this type of space link. If there is a need for emergency communications from a manned spacecraft or if a low-earth orbit satellite needs an urgent update of software or an emergency command, then an ISL link is essentially the only system that can establish instantaneous communications. While today's systems such as the NASA Tracking and Data Relay Satellite System (TDRSS) operates in the S and K-Bands in speeds measured in kilobits/second, the Advanced TDRSS and the European Data Relay System (EDRS) will operate in the second half of the 1990s in the megabits/second range. Eventually systems such as the hypothetical Asia/Pacific ISL link will operate in the gigabits/second range.

One of the several advantages of this type of services is that data on low earth orbiting satellites can be transmitted via ISL link directly to processing centers. Furthermore, since ISL links to geosynchronous satellites can be established from the low-orbit satellites wherever they may be above the earth's surface, the storage capabilities of these satellites do not need to be as great. Finally the developments in technology made in this application are generally transferable to other applications such as inter-regional ISL's

10.2.4: ISL Links Between Low Orbit Constellations of Communications Satellites

The first twenty years of satellite communications was focus on the use of geosynchronous satellites and to a less extent on highly elliptical orbits satellites for telecommunications services. This was because only a few satellites, say 3 or 4, were needed to establish global coverage. Furthermore, with geosynchronous satellites antennas could be continually pointed to the satellite and did not have to track the satellite except for occasional pointing corrections. Today, however, much different concepts for communications satellite systems for mobile communications and position determinations are being planned. These new mobile satellite systems (MSS) need much better "look angles" to the satellite to establish service to small hand held units that resemble cellular radio telephones.

The most effective concept to establish such systems is a global constellation of small satellites that blanket the earth under a network of 50 to 70 satellites. To ensure continuous connection among such a global constellation of satellites to support voice or services which do not tolerate store and forward operations, the use of intersatellite links is one available answer. In this case the lengths of the ISL are considerably reduced, but the relative speed of interconnecting satellites is greater and the number of interconnections a satellite might typically be completing increases from say one or two to four. There are at least two operational mobile satellite systems who currently intend to use ISL for low-orbit satellite connection to support system operation, network management, and some voice services. In general, these MSS systems will use the terrestrial network as the first routing option and use the limited capacity of the ISL connections only if all other available channels cannot provide the required service.

10.3 INTERSATELLITE LINKS TODAY AND TOMORROW

Although only a few satellite systems such as the Tracking and Data Relay Satellite and the Loutch Satellite are employing ISL today as an operational service, there are hundreds of millions of dollars being spent on research in this area and some six experimental satellite systems schedules for launch in the next four years with RF and optical ISL's on-board. There are today four ways to utilize ISL's for practical advantage. Adjacent satellite ISL allows one to add capacity to a satellite rather than replacing it. The adjacent satellite ISL is unique in that it could be accomplished by either physical fiber optic link or free space transmission.

The second option is the inter-regional ISL which can create new kinds of interconnectivities and avoid the delay and quality loss associated with double hop interconnection. This involves the longest ISL and perhaps the most difficult technological development. To date this application of ISL has not been accomplished in part because of the high opportunity cost associated with this service. This means that building and installing this type of ISL is very expensive. Furthermore it takes away mass and power from the satellite's main revenue generating capacity associated with space-to-earth and earth-to-space links. The net "value" added is such an ISL turns out to be less than the "value" lost for most operators of international satellite systems. As optical ISL technology improves the opportunity costs of inter-regional ISL will change and perhaps become economically viable in the 1990's.

The third option is the low-earth orbit to geosynchronous data relay satellite. This option is already extensively used and its utility will increase with high capacity optical ISL systems now being planned around the world. The technical developments made with type of ISL application will be increasingly applied in commercial uses of ISL for both low-earth orbit systems and inter-regional systems.

The fourth and final option in ISL use is that of low-earth orbit global constellation interconnect for mobile satellite services (MSS). The commercial viability of this type of ISL service will like stimulate the rapid development and innovative use of ISL in the 21st Century.

The unique ability of ISL's to exploit optical transmission above the interfering features of the earth's atmosphere as well as its omnipresent characteristic of linking any two points on or off the earth's atmosphere creates a tool of enormous power. With this type of technology "total" global communications without restraint or limitation becomes for the first time truly possible.

10.4 THE AMAZING NEW HALE PLATFORMS OF PROTO SPACE

During the last hundred years we have seen the advent of radio, microwave relay, television, submarine telephone cable, satellites, cellular radio telephone service, video text, fiber optic cable, and digital communications. Now we are on the verge of a remarkable new tetherless technology that will facilitate anytime, anywhere communications. What is remarkable about this new telecommunications service is that it is low in cost to implement, is very user friendly and is almost ideally suited to the needs of developing countries in terms of providing flexible coverage to both urban and rural areas with capital investments in the tens of billions of dollars rather than the hundreds of millions typically required for fiber optic cable or satellite systems.

This remarkable new technology, called HALE or the High Altitude Long Endurance platform, is very much like a geosynchronous satellite that flies in a very low orbit-. These 20 kilometer high communications platforms can provide the equivalent of wireless cable television directly to home receivers, or cellular mobile telephone service or rural and remote telecommunications or perhaps even a hybrid HALE platform than could do all three tasks at once. This is not to say that the HALE is a telecommunications cure-all that can solve every telecommunications problem. It is, however, a remarkable new development that is able to meet the urgent communications needs of many Latin American countries and that the capital investment to get started is modest in comparison to a conventional communications satellite project.

If you wish a quick understanding of the HALE platform, think of an "eternal remotely-piloted or robot high altitude platform" This HALE platform has been optimized to operate as a stationary communications tower. How did the HALE platform come to be and what are its future prospects? Read on and find out more about this latest innovation in the world's ongoing information revolution.

10.4.1 The Pre-History of HALE

To understand the potential of HALE technology and how it relates to the world of satellites some brief background is needed not only for space communications but telecommunications in general.

First we had the age of geosynchronous satellite in the 1960's. This began as a serious proposition when Early Bird was launched by INTELSAT in 1965. Suddenly we had global television, cheap reliable voice communications and South America was connected to the rest of the world as never before. As Santiago Astrain, former Director General of ENTEL Chile and the first Director General of INTELSAT once observed. "We in Chile were able to buy an earth station and link to the world by investing less than the cost of a jet passenger airliner. It seemed in the late 1960's almost a miracle. In a period of less than a decade a global satellite system was created and global telecommunications became a matter of routine."

By 1974 Algeria sought from INTELSAT the lease of satellite capacity for domestic television and long-distance national telephone and data service. A special rate for such domestic service was agreed and a network to connect each and every regional capital in Algeria was created in a short order of time. By the 1960's. more and more developing countries began leasing satellite capacity for domestic telecommunications services. About 50 countries lease satellite services today for their long distance voice and television services. A few such as Indonesia, India, Brazil, Turkey, Thailand, Korea, and Mexico proceeded to build and launch their own dedicated domestic satellite systems. The global satellite system became ever more extensive and ultimately linked some 200 countries and territories through more than 2000 international linkages. Some 50% of the overseas telecommunications travel by this means.

Beginning in the 1980's, however, two fundamental and interrelated changes occurred. First the restructure and divestiture of AT&T. This clearly was a move on the part of the US. government to promote competitive telecommunications service and to stimulate the creation of new telecommunications carriers and service providers inside the US.. Although this was less clear at the time it signaled that this new competitive environment would spread to many other countries.

Second, at the international level the same types of events also occurred. In particular, in 1984 competitive challenges were posed by private groups like Orion and PanAmSat to the INTELSAT global satellite system. Although INTELSAT was first envisioned as a single integrated satellite system and operated on this basis for two decades, it was by the mid 1980's being actively challenged by a number of new private satellite systems.

By the 1990's the demand for a broader range of satellite services, the desire to achieve access to lower cost competitive rates, the new digital communications technologies and smaller and lower cost VSAT ground terminals all served to create almost a torrent of change.

INTELSAT began offering INTELSAT Business Services and INTELSAT data networking services to antennas that were smaller and lower in cost. Some governments allowed more than one carrier to access the INTELSAT space segment to promote competition The other new private systems offered their own version of satellite services for video distribution and voice and data services. Some users of telecommunications such a video broadcasters attempted to use the new arrangements to by-pass the traditional PTT organizations around the world. Despite all these changes satellite technology remained expensive and largely beyond the reach of the more than 100 smaller countries whose geographic size and population patterns did not justify building a dedicated satellite system.

10.4.2: The Technology for Localized Broad Coverage Telecommunications - the HALE Platform

The fact is that satellite services are still relatively expensive. Especially dedicated satellite systems can easily cost hundreds of millions of dollars (US). Even the innovative and low cost small project that the Brazilian space agency INPE has recently designed, called Equatorial Circular Orbit -8 (ECO-8) will cost at least a quarter of a billion dollars and possibly up to $400 million (US). The problem is clear. The telecommunications needs are great around the world and the market is very attractive and growing, but the huge capital investment for both satellite systems and modern fiber optic networks is a barrier to rapid progress.

A few years ago, it was noted that satellite relays in the sky to provide telecommunications services really fly at a much higher altitude than is optimum for the communications function they provide. The best location for a telecommunications platform would in fact be much closer to the earth for the following reasons:

• Path loss. The further a signal travels through space the more power it loses. Furthermore is loss is calculated on the basis of "square" of the distance traveled.

• Frequency re-use. The closer the relay platform is to the ground, the greater the opportunity to use cellular technology to re-use frequency through creation of many cells.

• Transmission delay. The further the platform is positioned away from the users, the more time it takes to relay signals from the transmit and receive points. In the case of geosynchronous satellites the transmission lag is one quarter of a second one-way and a half a second round trip. For voice and data services this is considered a problem. (In the case of HALE platforms transmission delay would be under 10 milliseconds).

HALE platforms which are like very low orbit geosynchronous satellites enjoys all of these benefits. It still can provide coverage across a service diameter of 480 kilometer of 181,000 square kilometers and it involves only a small fraction of the cost of a conventional satellite regardless of whether the satellite is in low earth orbit, medium earth orbit or geosynchronous orbit. HALE's will operate in tandem pairs with one platform on the ground and one in high altitude. This means you can upgrade the capacity or payload of your system at any time and that you always have a spare available at all times. The flexibility of a HALE platform is considerable. A single HALE platform operating system can cover many dozens of countries around the world.

HALEs in short, are multi-purpose tools that can not only be used for every communications applications but could also be used for remote sensing, agricultural, forestry, and sea based monitoring, and other tasks as well. The platform can be held constant to within a one-half mile nominal location or it can be a free-flyer.

How then does the HALE platform work? There are in effect three key approaches to the powering and deployment of the platform itself. The first and most immediate way to put a HALE platform in position is with the use of high efficiency turbine engines which are many times more efficient than conventional airplane engines. These platforms are very lightweight and their is no human pilot so that the platform can devote the maximum amount of power and mass to the communications platform. These type of platforms would typically operate on a 7 to 10 day duty cycle, with one of the two tandem pair platforms always being aloft. This is the most available option today, but its operational costs are higher than the other options.

The second approach is the solar and fuel cell optimized HALE platform. This is also a very light weight platform that also resembles a high altitude airplane, but in this case solar cells provide enough power to keep the platform stationary during hours of sunlight. At night time high efficiency regenerative and unitized fuel cells provides the power to keep the electric turbines turning. This approach is very appealing in that it is environmentally very benign. This HALE platform essentially runs on sunlight and water. It can stay aloft 4 to 6 months at a time and thus its operating costs are less. This technology is projected to be commercially available in production quantities by the end of the century.

Finally, the third version of HALE propulsion systems involves using microwave or millimeter wave power that is beamed up from the ground. This approach could allow the use of one rather than a tandem pair of HALE platforms and would give almost perfect stabilization of the platform and perhaps at higher altitudes. This is the most exotic and long term concept and will not likely be commercially available for another 5 to 8 years. There is also concerns that all health standards considerations would be met with the microwave beam powered HALE. It is possible that the solar and microwave HALE platform could be combined in a hybrid system in 21st century operation.

The main thing is to note that the HALE platform offers strong service, quality and cost benefits to telecommunications users in both urban and rural settings. The HALE platform is certainly new, unusual, and at first view exotic. In fact, very little of the technology is truly original or unproved. Only the remotely piloted vehicle control mechanisms and the fuel cell technology is actually new and state of the art. The actual platform design and construction techniques could in theory be transferred to countries wishing to establish capabilities in this area. The antennas to be used on HALE platforms could in fact vary a great deal in sophistication. Early design seeking state of the art performance has examine some very sophisticated multi-cell phase array antenna systems that could allow one HALE platform to offer a 50 to 100 digital television entertainment and educational channels or up to 15,000 simultaneous cellular telephone channels. It would be possible to use much less demanding communications technology say for rural and remote service.

The HALE technology of the next few years should offer a wide range of communications or even remote sensing capabilities. Before this occurs, however, a number of questions will need to be answered. Prime among these questions are the following:

• How much will HALE platforms cost? It is still to early to obtain specific prices and there is question as to whether one would want to "buy" or "lease" the platform in order to obtain the most reliable and cost effective capability. A rough order of magnitude of the cost of the more conventional HALE platform would be about $5 to $6 million per copy.

• When will HALE platforms be available for purchase or lease? It is currently anticipated that the piston-based versions will be available in 1997 and the solar and fuel cell versions will be available by the end of the century.

• What about the issues of frequency allocations, licensing, access to local telecommunications, radio or television markets, etc? The plan would likely be to make HALE technology available to national governments or authorized telecommunications carriers. In short, they are not seeking to provide services, except in partnership with local entities. The most likely scenario is to provide the technology to those with appropriate licenses and access national markets through local officials and agencies. There is a possibility that HALE technology might be combined with a low earth orbit satellite system to augment that systems coverage and cost efficiency. In such instance, then the satellite carrier would need to obtain all necessary regulatory approvals.

• Just how cost efficient are HALE platforms in comparison to other space or terrestrial options? The answer to this of course depends upon the area of coverage and the populations within the target service zone. For the optimum service radius of 240 kilometers, however, the HALE platform has a cost benefit advantage of towers of 4-to-1. Furthermore it is 15-to-1 over cellular tower systems, and an incredible 22-to-1 over satellites for this zone of coverage.

It would be easy given the excitement, newness, and major new multi-purpose capabilities of HALE platforms to oversell and overemphasize the importance of this new development. The truth is that HALE platforms, although truly exciting, simply give us yet another important telecommunications capability—some of which are uniquely well suited to developing countries with special needs. No one should, however, believe that this can or will meet all telecommunications needs on a universal basis. It is thus a key but not a comprehensive tool that fits well between terrestrial towers and satellites. In many cases HALE platforms will work best as part of an integrated plan that involves terrestrial fiber, satellites, and other existing systems.

10.5 CONCLUSION

The growing acceptance and implementation of inter—satellite links is a result of the maturation of ISL technology using both RF and optical systems as well as market demand. The ability to provide direct connectivity and the flexibility to spread traffic among available system capacity. Up to this point in the development of the satellite industry, the "opportunity" costs of ISLs in comparison to more communications capacity were high. As ISL technology has matured with performance increasing and costs declined, the economic attractiveness of this concept has increased. Meanwhile the system flexibility of ISLs have also increased in value in terms of enhanced network performance. As a result ISL will found in future systems from LEO to GEO networks.

The other technology that has the potential to extend the capacity and performance of satellite systems and also creates a whole new category of "proto-space" telecommunications services is the so-called autonomous platform. These High Altitude Long Endurance (HALE) platforms will create a whole new capability which has a huge potential that is only beginning to emerge. The new frontier represented by proto-space can be effectively and cheaply used for not only telecommunications but also news gathering, remote sensing, and many other uses. The HALE platforms give promise of a new multi-billion industry that may well link up with satellites to form new and innovative hybrid systems. The one thing about satellite services and technology is there is never a boring period in its continuing and fast-paced evolution.

CHAPTER 11

LEOs AND MEOs:
THE NEW SATELLITE SYSTEMS ON THE BLOCK

"The National Information Infrastructure and the Global Information Infrastructure have been compared to the railroads and the interstate highway system, but these are static models of the future. We need to think of 'off-the-road vehicles' that can jump the tracks and bring communications wherever they are needed. We must think of flexible links that can connect to anywhere at anytime to anybody through mobile satellite and wireless systems available throughout the world."

Craig McCaw

McCaw Communications

11.0 LOW AND MEDIUM EARTH ORBIT SATELLITE SYSTEMS

For 30 years there has been rather continuous technological innovation within the field of geosynchronous communications satellites. During the last few years, however, a revolution in technology has served to create a new era of change, re-focus, and re-conception to the entire field. The forces of competition vis a vis fiber optic cable plus new technological innovations in space communications are giving rise to new types of architecture, new orbits and total systems design.

The characteristics of these new systems include low latency and echo, low path loss, improved elevation angles, high latitude coverage, high levels of frequency reuse, and the prospect for improved quality of service, especially for mobile services. To date these new systems as formally proposed can be grouped as follows: (a) Little LEOs, operating at 800 MHz or below; (b) Big LEOs and MEOs operating at 2 GHz or above; (c) and Mega-LEOs which could operate at various frequencies but have so far been slated for 30/20 GHz. Finally, the most recent concept which is just being commercialized is the new type of telecommunications platforms which could maintain geosynchronous operation in what has been called proto-space. These would called High Altitude Long Endurance (HALE) platforms and could presumably offer a broad range of services in a variety of frequency bands. This will discussed, however, later in this chapter as a separate technology.

The basic characteristics of the various types of new satellite systems that have been proposed are summarized below along with some attempt to contrast and compare their capabilities.

Since no major commercial, near-earth orbit systems for mobile voice service now exist, we will have to use an analytic framework to show the many trade-offs that various types of systems can achieve in terms of coverage, look-angles, frequency re-use, path loss, operational performance and net operating and system costs. This will allow a type of comparative algorithm to be developed to present not only the individual characteristics of all proposed systems but also present a comparative figure of merit for all types of systems including geosynchronous, medium earth orbit, big and little LEOs, and mega-LEOs.

The key to applying this comparative algorithm is the relative weighting that one applies to the various factors used to make the evaluation. These include such factors as latency, coverage of higher latitudes, levels of frequency reuse, cost, size and portability of ground transceivers, etc. In short, this comparative analysis attempts to apply reasonable and fair assessment factors, but its conclusions can be altered by placing different priorities and weightings as to which service characteristic is considered most important.

Ultimately, the demand for services and the marketplace will decide which systems are funded, built, deployed, and operated successfully. In the meantime, this analysis does attempt to present in a comprehensive matter all systems now envisioned and the various strengths and weaknesses of these systems on a comparative basis. Clearly a satellite system designed for fixed satellite service, for VSAT service, for personal communications services, or for Direct Broadcast Service will have different figures of merit and the comparative index developed within this article can be best applied in comparing systems designed for a particular service rather than systems with totally disparate service objectives. Despite these limitations, it is hoped that the comparative methodology presented here provides at least a better way of understanding ways in which the proposed new systems both differ and compare to traditional systems.

At the risk of oversimplification, the areas where fixed satellite technological advancement have produced the biggest performance increases and the largest cost reductions are in the following areas: (a) increased on-board power; (b) increased frequency reuse through multi-beam antennas and cross polarization;

(c) constantly focused spot beam antennas; (d) digital modulation, encoding and compression techniques; (e) beam switching and interconnection; and (f) lifetime extension techniques. Clearly other factors have produced some gains and helped to simplify earth station antenna design, but the five factors listed above were most key. These factors have all combined to make geosynchronous satellite systems over a thousand fold more cost-effective than the satellites of the mid 1960's. These innovations also allowed contemporary satellites to offer a higher quality signal as well.

In most industries, an technology which achieved a 1000 fold increase in productivity or cost efficiency over a 30 year period would be considered truly phenomenal. In the fields of telecommunications and computers, this rate of innovation in terms of the so-called Moore's law is considered normal or expected. (Moore's Law predicts a doubling of performance every 18 months.)

The parallel development of fiber optic cable systems and digital processing has indeed applied competitive pressure to the satellite field. The industry has also beginning to evolve in new directions as the potential of new markets and services began to be better understood. In order to compare new satellite systems with conventional geosynchronous satellite systems and with fiber optics we need to first devise some criteria for a score card.

11.1 SATELLITE SYSTEM CHARACTERISTICS AND TRADE-OFFS

The key parameters for a satellite system design (or any telecommunications system for that matter) are nearly infinite. The principal options, however, can be represented by some twelve features which although large are still a practical number to be considered and comparatively evaluated. These factors for evaluating one satellite system against another (i.e. a score card) are as follows:

- Connectivity
- Cost and complexity of ground transceivers.
- Cost and complexity of satellites
- Ease or difficulty of achieving frequency re-use
- Efficiency of coverage of prime coverage areas
- Ground antenna tracking and gain requirements.
- Launch cost and launch options
- Lifetime
- Look angle to satellite
- Number of satellites needed to achieve coverage
- Operational requirements for TTC&M and autonomous control
- Path loss
- Propagation time, transmission delay or latency

- Requirements for a Inter satellite Link
- Sparing strategies and concepts

The relative importance of these factors is dependent upon many other factors such as the type of service to be provided, the frequency band to be used, and the patterns and configurations of intended users. This is why the factors are listed in alphabetic order. The most likely or often way of determining a relative figure of merit is a calculation of cost of service that is somehow figured on a parallel or common basis. The chart in figure 30 essentially seeks to calculate the cost of comparable satellite service delivery (i.e. Megabits per second) over the service area covered. This is certainly not a perfect measure of a figure of merit. Certainly, it gives advantage to those systems with higher cost and more elaborate ground systems and less advantage to those with lower cost ground systems. Nevertheless it is still an interesting and useful measure.

This metric as presented in figure 30 suggests that geosynchronous satellites are highly cost effective for large coverage areas and conventional fixed point-to-point satellite services with concentrated traffic and a more limited number of earth station facilities. If one adds a very large number of users with the need for lower cost and mobile terminals then the various LEO and MEO options will start to look increasingly better. Perhaps the biggest surprise of all is that the new concept of a High Altitude Long Endurance Platform seems to hold the advantage of coverage, throughput, look angle, and cost per area served without sacrificing the ability to provide low cost earth segment facilities as well.

Clearly these comparison are at this stage highly theoretical since GEO systems exist while Big LEOs, Mega-LEOs, MEOs and HALE Platforms are only proposed. One of several key unknowns in all this is the importance of latency and transmission delay. Many believe that in the 21st century satellite technology must reduce delay significantly if it is to stay competitive with fiber optic technology. If this is indeed true then low earth orbit satellites, whether they are global constellations or equatorial circular orbits or some other configuration will be increasingly important. Likewise HALE technology will likewise increase in importance and market share.

The complexity of satellite typology has increased greatly in the last few years at least in terms of planned satellite systems. The advantages of such satellite systems will actually be demonstrated as systems are deployed and are tested in the real world market. In the interim some type of comparative chart that indicates the relative strength and weaknesses of the various approaches. Such a chart is provided in figure 31. This chart does not attempt to develop a comparative single metric as that presented in figure 30, but rather it seeks to show within a single chart the strengths and weaknesses of various types of

FIGURE 30

COMPARING SATELLITE SYSTEM CAPABILITIES

SYSTEM TYPE	SYSTEM COST ($US)	COVERAGE (SQ. KM)	BEAM THROUGHPUT	BEAM PERFORM. INDEX
GEO (3 Sats) 8 Beams) 7 YR LIFE	$1.2 B	12.6x 10	200 Mbs	$2,834/Mbs/Km^2 per year
MEO (12 Sats) (5 YR LIFE) 20 Beams	$2 B	3.1×10^6	250 Mbs	$6,451/Mbs/Km^2 per year
LEO (50 Sats) 40 Beams) (5 YR LIFE)	$4 B	1.5×10^6	50 Mbs	$8680/Mbs/Km^2 per year
MEGALEO (800Sats- 50 Beams) (5 YR LIFE)	$12 B	1.5×10^6	100 Mbs	$6000/Mbs/Km^2 per year
HALE (12 Cells) (10 YR LIFE)	$10 M	0.18×10^6	200 Mbs	$1852/Mbs/Km^2 per year
TERRESTRIAL (30 Cells) (20 YR LIFE)	$50M	$.008 \times 10^6$	90 Mbs	$82,000/Mbs/Km^2 per year

SAMPLE CALCULATION:

GEO SYSTEM:Costs ($1.2 Billion) divided by: (200Mbs) X Beam size(12.6 M.Km) X Sats(3) X beams(8) X life(7 yrs) X (eff)100%

Resulting Beam Performance Index: $2834/Mbs/Sq. Km./Yr

Note: Efficiency index for MEOs, LEOs, and MEOs is 30% because of ocean and artic coverage 70% of the time.

Note: The HALE platform cost assumes two platforms of $5 million each continuously providing service on a rotational cycle each lasting 4 months duration.

FIGURE 31

COMPARISON OF DIFFERENT TYPES OF SATELLITE SYSTEMS CAPABLE OF REAL TIME SERVICES-VOICE AND ABOVE

Type of System	Number of Satellites	Freq. Re-use Opportunity	Xmit Delay	Path Loss	Look Angle	Ground Segment
GEOSYN	few (3)	low (4 to 6 times)	high (0.25 sec)	high	low to high	typically expen. to med.
MEO	medium (12 to 18)	medium (8-12 times)	medium (> 0.1 sec)	medium to high	medium	medium expen. to low
LEO	high (50 up)	high (30-100 times)	low (>0.05 sec)	low	high	low
MEGA LEO	very high (hundreds)	high (30-100 times)	low (>0.05 sec)	low	high	ranges low to med
Equator Cir. Orb.	typically med. (8-15)	medium (8-12 times)	medium (> 0.10)	medium to high	medium	ranges low to med
HALE * Platform	one per region	medium (8 to 12 times)	almost none (>0.001 sec)	almost none	high	low

* NOTE: High Altitude Long Endurance Platforms operate at altitudes of 19 to 24 kilometers and are in what has been characterized as proto-space or the 18 to 110 kilometer region of space. They typically can cover a region with 480 kilometer diameter or 180,000 sq. kilometers. This is included for providing a wider range of comparisons.

satellite systems as well as that of the HALE platform which operate not in outer space but so called proto-space.

11.2 THE RELATIVE MERITS OF DIFFERENT SATELLITE ARCHITEC-TURES

The field of satellite communications is currently in a period of major transition. This transition is being driven by the development of a wide range of new services and applications that includes direct broadcast satellite services, all forms of land, air and sea mobile communications, plus navigational and radio determination services. This is in addition to the growth and development of conventional geosynchronous fixed satellite services. The field of satellite communications is likewise being strongly influenced by a range of new technologies, system architectures and satellite orbital configurations.

The cumulative effect is that the satellite communications field is in a period of rapid and highly dynamic change in almost every regard. These changes are perhaps most driven by the use of new orbital configuration to achieve better cost or technical performance. Other factors which will be addressed later in the Chapter will also have a major impact on market success.

Is it therefore possible by examining orbital configuration and planned technology to predict which satellite systems of the next decade will succeed and which will fail? Obviously, the answer is no. First of all the key to success in the satellite field is first and foremost rooted to market factors. Those with extremely effective market ties and which are well established systems such as INTELSAT and INMARSAT have a great advantage that transcends technological and operational issues. Secondly, the various approaches to deploying satellite systems that are characteristic of many of the proposed satellite system have widely different technical, operational and market factors behind their designs.

11.3 GUIDE TO THE NEW SATELLITE SYSTEMS

The following review thus examines the rationale behind the various MEO, LEO, mega-LEO and ECO systems that have been filed in the US. or internationally. This analysis assesses the logic behind the systems using the comparative methodologies found in figure 32 below. There are so many subjective or non-comparable factors involved that this analysis cannot be considered scientific or even quantitative. Nevertheless the differences in the systems can be shown in rather clear relief and the strengths and weaknesses made more clearly identifiable.

Figure 32

Review of New Satellite Systems Providing Voice and Above Service

System Name	System Type	# of Sats	Basic Rationale
ECO - 8	ECO	8	Use of a series of low to medium earth orbit equatorial circular orbit or equatorial zone satellite system to provide continuous global coverage of the equatorial zone. Developed by INPE, the Brazilian Space Agency to cover not only Brazil but other equatorial countries as well. Relatively low cost, but it will be difficult for a nationally based system to sell service to other equatorial countries. **Key Constraints:** Raising capital needed to complete project. Limited frequency. Essentially limited to voice and data. Limited spectrum.
Ellipso	Elliptical	N/A	Use of a number of elliptical orbit satellites to concentrate coverage over North America. This medium earth orbit elliptical satellite system thus places maximum capacity over key targeted market areas. This is a rather straightforward bent-pipe design that will be able to provide satellite cellular service in the $0.50 per minute and up range. **Key Constraints:** Raising capital for a start-up enterprise. Competing with other satellite cellular systems which are more well known plus competing with terrestrial cellular as well. Limited spectrum at 2 GHz.

System Name	System Type	# of Sats	Basic Rationale
Globalstar	LEO	50	Use of a LEO constellation to provide world wide cellular telephone and data services. Use of a simpler"bent pipe" design without ISL to simplify system and make it much lower cost. Concept is to deliver a basic cellular telephone service at about $0.50 per minute or at one-eighth the cost of Iridium. Also uses CDMA spread spectrum techniques to boost capacity since Qualcomm, one of the partners in Globalstar, holds leading position in this technology. Also Globalstar has international backing as a US-French venture with major investment by ALCATEL through the Loral Corporation. **Key Constraints:** Limited spectrum at 2 GHz. Market appears to have too many entrants. Limited value-added service offerings
INMARSAT	MEO	N/A	Use of medium earth orbit constellation in conjunction with existing INMARSAT geosynchronous constellation. Existing traffic base in aeronautical and especially maritime mobile service gives major head start. New Project 21 will, however, be semi-autonomous. Limited number of satellites, perhaps 12 or less, will reduce capital investment in comparison to say Iridium system. **Key Constraints:** Medium earth orbit will have more delay, more path loss, and some lower look angles. Satellites will need to be more complex and more powerful to

System Name	System Type	# of Sats	Basic Rationale
INMARSAT	MEO	N/A	compensate for this. Limited spectrum as 2 GHz. Price must be competitive will market yet not undercut existing maritime and aeronautical service.
Iridium	LEO	66	Use of a LEO constellation with ISL to achieve the following: (a) global coverage; (b) mobile communications with hand-held transceiver; (c) 48 fold frequency re-use. (d) executive mobility land mobile service anywhere. (e) bypass of terrestrial network when needed. (f) first in to the market Pioneer's preference (denied) (g) key profitability to be in hand-held units. (h) global consortium to finance $4 billion space segment. **Key Constraints:** High cost of space segment and high cost of service i.e. $3 to $4 per minute.
Odyssey	LEO	12	Use of a medium earth orbit system to provide global cellular voice and data services. Reduce the number of satellites required by going to higher orbit. This reduces opportunity for frequency reuse, increases path loss and alters look angels, but overall allows a low cost service, projected to be under $ 0.50 per minute. **Key Constraints:** Despite very good technical design, there has been difficulty raising the capital for this project by its key proponent TRW Inc.
Spaceway	GEO	8	Use of highly efficient cellular spot beam technology and the Ka-Band frequencies to attempt to provide a global system with the range of service offerings comparable to a Teledesic say a mega LEO system. This system suggests that if GEO systems push the technology as far as possible they

System Name	System Type	# of Sats	Basic Rationale
Spaceway	GEO	8	can provide a cost-effective and competitive service and do so for mobile hand-held transceivers, for village based rural telecommunications systems and even high speed broad band services. Seeks to deliver service at under $0.25 per minute. **Key Constraints:** Reliance of Ka-Band technology which has yet to commercially proven despite experiments (e.g. ACTS, ETS-6). Latency associated with geosynchronous transmission. Very large capital investment in unproved markets.
Teledesic	Mega	840	Use of a huge number of mass produced one-ton but very high performance satellites with more than 50 fold frequency reuse in the Ka-Band. Unique technologies with up to 8 fold ISL connections, fixed or "painted spot beams", super computer in a "shoe box", etc. Based on conversion of US. military "brilliant pebbles" program. Also envisions commercial use of Ka-Band Seeks to provide service at under $0.25 per minute. Projected cost of system estimated at $12 billion or more, making it the most expensive and largest capacity of any system. **Key Constraints:** Even with the backing of Microsoft's Bill Gates and Greg McCaw, the huge capital needed for such an exotic and unproved design would seem very difficult. Marketing and sales for such a huge system on a global scale would also seem to be a major challenge.

Clearly not all of these proposed systems will succeed. Many major players in global telecommunications have not yet decided how they will proceed in this sector. There are clearly opportunities for mergers and acquisitions among the systems described above. The commercial and market driven aspects of this field are, in fact, likely to have as much if not more impact than the rapidly evolving technology. Nevertheless, some conclusions are possible at this time.

The first conclusion is that satellite technology will continue to evolve rapidly among all of the system types described above and that this evolution will allow satellites to remain competitive with fiber optic systems in at least certain key markets and applications.

The second conclusion is that the market will largely determine the success or failure of LEO, MEO, Elliptical, and ECO satellite systems over the next 5 to 10 years and that cost efficiency will be only one of several indices of success with quality, coverage, latency and user friendliness being other key factors,

The third conclusion is that institutional arrangements, global financing, landing rights and liberalization of telecommunications systems at the national level will be the other key factors in the successful deployment of the new satellite systems. Today the virtual domination of US. firms in this area could serve as a major factor in attempts to block such systems until a higher level of international and non-US. participation is achieved. Clearly this dynamic evolution will be extremely interesting and eventful with a host of technological, political, regulatory, and financial milestones yet to come over the next decade. The added dimension of the new High Altitude Long Endurance (HALE) will only add to the complexity of this rapidly changing space communications environment.

11.4 CONCLUSION

There is no right or wrong in terms of satellite system design or orbital configuration in today's competitive market, the right design is the one that delivers reliable and high quality services and affordable market rates. The expansion of service requirements to cover land, maritime and aeronautical mobile services, broadcasting, computer networking, fixed communications services, navigational and positioning services, and a host of other factors is driving systems toward greater technological and service diversity. Especially the needs of mobile satellite services as well as broadcasting has help to develop new classes of satellites, new orbital constellations and new modes of operation. Messaging service alone has given rise to a whole new class of store and forward small satellites. This burgeoning and expanding market in space communications can thus be expected to give rise to an even richer mix of satellite types in the 21st Century.

══ CHAPTER 12 ══

THE FUTURE OF SATELLITE COMMUNICATIONS

> **"Prediction is difficult, especially about the future."**

Edgar Fielder

12.0 INTRODUCTION

The world's first artificial satellite, SPUTNIK, was launched on 4 October 1957. The first communications satellite SCORE came in December 1958, and was capable of broadcasting only a single voice taped message by President Eisenhower "Peace on Earth, Goodwill to Men". By 1965, when the world's first commercially operational satellite, Early Bird (or INTELSAT I) was launched, the capacity had expanded to 240 voice circuits or a single "grainy" black and white TV channel. A quarter century later, the INTELSAT 700 series satellites represents today's state-of-art for space communications and the 800 series is under development. These satellites represents an awesome capability that totally dwarfs the Early Bird satellite. The INTELSAT 700 series can transmit 5.1 billion bits of information per second, sufficient capacity to send the entire Encyclopedia Britannica plus color facsimile of all illustrations in under 2 seconds. Although there is little demand for such service, it is still an impressive statistic.

Today, an INTELSAT 700 series satellite can send 56,000 analog voice circuits at once, but with the new digital techniques available its capacity has expanded enormously. It will almost be like blowing up a balloon.

Thus, in thirty years, the capacity and performance of satellites have zoomed forward with amazing speed. Only the computer industry has moved ahead with greater and more rapid progress. From the days of SCORE and Courier 1B is a journey of just over 30 years, but the difference is astonishingly great. It is like literally comparing the height of an ant and the World Trade Center in New York.

With such tremendous progress for a sustained period of thirty years, there is a temptation to assume you can extrapolate the future from such a steady rate of growth in both technology and performance. This is far from true.

There is a rule in the prediction business. It goes like this: "Don't make a 5-year projection or prediction because you will be too conservative. Most likely you will not anticipate a major new technical development. So make a 25 to 50-year prediction. You may not be correct, but there will be few people around to check up on you".

12.1 THE "ENVIRONMENTAL" FACTORS

Regardless of which time scale is used, projecting the future of space communications accurately is certainly difficult. The "environmental" factors which complicate accurate prediction preclude the following:-

- Fiber Optic Cables: The rapid growth and technological progress of fiber optic cable systems as well as their increasing cost efficiency must be recognized. In short, cables are now a major technological competitor to satellite, especially at the national level, but also increasingly at the international level as well.

- Setbacks in the Launch Vehicle Industry: The lack of truly reliable and truly cost effective launch services is, today, perhaps the single most important deterrent to the future development of satellite communications. The new privatized launch vehicles being developed in the US., and the increasingly sophisticated Japanese launch capabilities give new hope for the future.

  In particular the Pegasus launch vehicle now seems poised to support a number of new low-orbit satellite systems such as the new Motorola Iridium network. For every Pegasus development which achieves a breakthrough, there seems to be a large number of unsuccessful launcher projects in terms of achieving limited or no gains in cost efficiencies.

- Major Regulatory Shifts: The world of satellite communications is currently in the midst of regulatory change and institutional reform. Privatization or deregulation is under way at some stage in dozens of countries. A major impact is unavoidable as the new competitive systems such as ASTRA, ASIASAT, PANAMSAT and ORION will capture market share.

- New Patterns of User Demands: The patterns of use of satellites are clearly in a state of transition. HDTV and Video; all forms of digital services, especially ISDN-related; thin-route services to personal communicators, all forms of mobile services, and even fiber optic cable restoration will play increasing roles, while plain old telephone service (POTS) will diminish in influence.

- <u>New Technical Breakthroughs</u>: The satellite industry has largely been extrapolating known technology for over a decade. Digital circuit multiplication is the major source of progress in terms of improved cost-efficiency. There is, however, indication that satellites could be on the verge of a series of new technological breakthrough in the Twenty-first Century such as effective use of super-conductivity in space, "intelligent" low orbit satellites, and satellites with near-zero second delay stabilized in orbit by ground transmitted microwave or laser power.

Despite the difficulties to predicting the future of satellites we will break all the rules and try to anticipate the future 5, 10 and even 25 to 50 years from now. Our only guide will be to recall the words of futurist Edgar Fiedler, who said, "It is very difficult to forecast, especially about the future".

12.2 EVOLUTIONARY CHANGES

We can perhaps best start with a look to historical trends. The development of satellite technology from say, INTELSAT I to INTELSAT VI shows the following:

- <u>Antenna Development</u>: Antennas have evolved from simple low-gain omni antennas weighing less than half a kilogram to large complex antenna systems that can consume 20 to 25 percent of the mass of the satellite. These new antenna concentrate power thousands of times more efficiently than in the past and allow frequencies to be reused six times or more through polarization and beam separation.

- <u>Power</u>: The power systems of satellite systems have increased in capacity on the order of 100 to 1,000 times, through higher performance/higher efficiency batteries and solar cells, through much larger solar array systems, and larger power systems in terms of mass and size.

- <u>Higher Frequency Bands and Frequency Reuse</u>: The satellite communications industry has evolved in the direction of using much higher frequencies in the microwave region (i.e. the Ku and Ka-Bands), as well as the more intensive reuse of the lower bands (i.e. the C-Band). This was essential since the service requirements for space communications has increased more than 1,000 times in the last 25 years.

- Microelectronics and Digital Communications: While power and antenna systems have increased in size, volume and mass, the electronic components of the satellite have become smaller, less massive and more reliable largely due to the use of solid-state electronics. As space communications have shifted away from analog communications and in the direction of digital communications, it has allowed electronic performance to increase while the components have shrunk in size. In many ways satellites have become very high speed digital processors — super computers in orbit.

The question is will there be a continuation of steady evolutionary and gradual change in antenna performance, greater power, improved microelectronics and higher frequency use? Further, does such a future also imply we will continue to see smaller lower cost earth station antennas and receivers more accessible to users? The answer is "yes ... but we will also see what might be considered radical new departures in satellite transceivers as well."

12.3 REVOLUTIONARY CHANGE FACTORS

First let's consider whether there are any revolutionary, new technologies that could change space communications in a major, step function way. Between now and the year 2000 this seems unlikely, but by the Twenty-first Century new stabilization and launch systems, inter satellite links on board procession, phase array multi-beam antenna systems, artificial intelligence, super conductivity and advanced procession techniques should make possible revolutionary change.

Cost of service could drop to less than 10 percent of today's charges (adjusted for inflation, of course). A wide range of new services could also evolve. These include, direct broadcast satellite services (DBS); High Definition Television distribution, a wide range of new ISDN services especially for telemarketing and remote back-office operations. We will also see more accessible mobile communications including hand-held or briefcase equipped "satellite telephones".

There will also be totally new service such as multiple rastered television or some other form of three dimensional television or a scaled down version of holovision. Full scale holovision, designed for a 360° field of view vertically and horizontally would of course require awesome transmission rates on the order of 1 trillion bits per second and this is clearly many decades away. The 3D-TV that develops within the next 20-25 years will operate at speeds at least 100 times less than a terabit per second.

By the early to mid-Twenty-first Century digital satellite transmission capabilities in the 10-100 Gigabit per second range however, may well be achieved. This is the key for satellite communications since fiber optic transmission using solution pulses will undoubtedly achieve these rates. A terabit per second either by fiber optic cable or satellite is probably several decades away in terms of usable, cost-efficient technology. In short, an evolutionary period from now to the Twenty-first Century, will perhaps be followed by an exciting period of potentially revolutionary changes in the 2000 to 2020 time period. This period, which will roughly coincide with the colonizing of the moon will be a period of widespread change and exciting developments for humankind.

12.4 PREDICTING THE FUTURE

The satellite communications industry is second only to the computer industry in terms of rapid innovation and technological evolution. When we talk of "evolutionary" change involving 10 to 20 percent gain per annum in traffic and in technological performance, it may seem "revolutionary" to industries where 3 to 5 percent gains are considered normal or even significant. Satellite communication now some 30 years young still has enormous technological dynamism. Change that now seems likely for the future includes the following:

12.4.1 Satellite Antennas

As indicated earlier in this Chapter, this has to be an area of dramatic gain. Over the last thirty years antenna performance has increased on the order of one thousand to one million times. Very low gain omni antennas have given way to 3-axis stabilized spot beam antennas and now offset multi-feed and multi-beam antennas with polarization discrimination. This type of antenna technology has already produced multiple re-use of frequencies and down-link power levels of 65 dBW. (This is sufficiently high power to receive a high quality TV signal using only a 40 cm. diameter antenna which has a surface area of less than two square feet).

Antenna development, however, has an almost open-ended future. We may well see multi-beam antenna with hundreds of feed elements capable of creating 20 or more precise spot beams. Today's beams as shown in Figure 33 are very crude and imprecise and tend to waste precious bandwidth and power. This new antenna technology will likely use a single large reflector that allows perhaps dozens of reuses of the same frequency. This increase from today's six-fold reuse to tomorrow dozens should certainly help relieve the problems of geosynchronous orbital arc congestion. To build large reflectors cheaply and reliably, unfurlable antennas such as was demonstrated by ATS-6 may make a reappearance. Furthermore, flat "unfoldable" phase array irradiators with

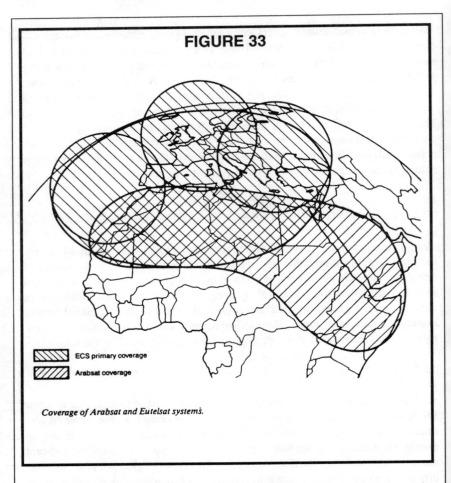

FIGURE 33

ECS primary coverage

Arabsat coverage

Coverage of Arabsat and Eutelsat systems.

electronic steering and ground based reprogramming of beams may co-exist with more "conventional" multi-beam antennas in the Twenty-first Century.

These so called "flat phase" array antennas are beginning to enter the earth station market in the US and Japan and soon Europe. They will undoubtedly migrate from ground-based use to space-base use.

Their advantages are clear. They do not need to be perfectly shaped and formed as single units like parabolic antennas. There is almost infinite flexibility to reprogram the micro electronic elements in the array to create new beams and coverage or demands. Further, there is no problem with unwanted side-lobe characteristics with the beams which can be almost perfectly formed and shaped.

This new phase array technology raises many questions. Prime among the unanswered question is whether should be used with clusters of small satellites, large platforms or even both.

12.4.2 Power

Looking at a 30-year time horizon and how space communications will be powered over that time scale is a challenging exercise. While antenna developments seem likely to evolve from known technology, power has lots of totally new options.

These options, in order of increasingly exotic technologies are as follows: (a) New and improved power generations on board geosynchronous satellites (e.g. improved multi-junction solar cells), solar concentrators, all battery powered satellites using advanced batteries, solar cell arrays with better solar orientation, thermionic converters; (b) Supply of energy to communications satellites from space-based satellite solar power systems using laser or microwave relay (perhaps 10KW on up); (c) Use of ground-based energy systems to stabilize a satellite at a fixed location about 500 miles in altitude and use the same system to supply all satellite power needs.

This much is clear, the technologies under option (a) could and likely will be developed within the next 10 to 15 years, while options (b) and (c) are 15 to 30 years out in terms of possible implementation.

What seems most clear is that requirements for power are still on the up-escalator. Interactive DBS satellites that provide to the home satellite service plus a return channel to the satellite may lead the way, but more likely mobile services to hand-held beepers, and briefcase satellite telephones will drive power requirements up from today's 1KW systems to at least 10 KW or even 20 KW systems in the Twenty-first Century.

There's one joker in the deck as far as the evolution power is concerned. This is the concern for reliability and security. Military programs and civilian ones too will probably move from silicon to gallium arsenide solar cells and perhaps, other high valence materials, not just because of increased efficiency, but also because of greater security from radiation damage.

12.4.3 Higher Frequency

The current trend in satellite communications if any is to back away from higher frequencies (i.e. the Ka-Band or the 30/20 GHz frequency range). Only the Japan space organization (NASDA) and Japanese industry (particularly the Japanese computer company, Fujitsu) is truly placing high emphasis on Ka-

Band. This is because terrestrial microwave usage in Japan in both C-Band (6/4 GHz) and Ku-Band (14/12 GHz) is already extensively used.

Further Fujitsu sees a strategic advantage in offering direct to the office computer service via Ka-Band satellite. In the US, AT&T and others have filed for 30/20 GHz satellites, but their timetable to implement is not clear. In short, more emphasis is being placed on using C-Band and Ku-Band with higher efficiency through digital circuit multiplication equipment or multiple frequency reuse.

With the development of NASA's ACTS satellite, the European Space Agency own 20/30 GHz satellite Olympus (now out of service) and Japan's innovative experimental satellite program, especially the ETS V1, the late 1990's will likely see a resurgence of interest in Ka-Band (30/20 GHz) and above.

In fact, by the time the Twenty-first Century is well under way significant moves to use millimeter wave frequencies and even light wave frequencies for space communications will become increasingly common. New technologies, such as superconductivity in the natural space environment, optical switching and the super speed requirements of parallel processors, will all serve to make the use of higher frequencies a natural fallout of Twenty-first Century space communications.

All of these developments: satellite antennas, power and frequencies, will allow steady progress. Advances in digital communications will certainly help to keep the pace of telecommunications moving ahead at a good clip. All in all though — expect no miracles. The miracles are not likely to be along until about 2010.

12.5 THE LONGER TERM FUTURE

The first 15 years of the Twenty-first Century should see a renaissance of satellite communications. Contemporary suggestions that fiber optic cables will virtually replace all satellite applications by the Twenty-first Century rolls around will seem silly if not quaint by that time.

The current way of launching satellites into geosynchronous orbit are awfully inefficient. Sometimes a 100,000 pounds of fuel and machinery are consumed to get one pound of payload into the proper orbit. By 2010 we may have learned fundamentally different ways to accomplish our goals. Here briefly, are some speculations on what might be called "radical options" for the future.

12.5.1 SPIDERSAT

SPIDERSAT would be a "near zero second" satellite stabilized in low earth orbit by ground based power. These very light-weight satellites that consist largely of membranous antennas would be raised into orbit and "stabilized" at about 500 miles by mean of say 4 to 6 microwave or photon transmitters. These transmitters would "launch" the satellite; stabilize it in a constance position and also supply the necessary power for its operation.

The SPIDERSAT would need not thrusters, no batteries and no solar cells. It would be all antennas and communications electronics. Once the ground-based microwave transmitters were built launches would be very cheap. Incidentally, this would also "solve" the satellite transmission delay problem and the transmission "path loss". Both concerns would be greatly reduced in the satellite altitude were lowered from 22,237 miles to 500 miles.

12.5.2 Tethered Satellites and the Pelton Space Escalator

Super tensile strength tethers are being developed which can do amazing things. Their tensile characteristics start to rival that of diamond. Tapered cables that start out a meter in diameter and gradually reduce in size to a few molecules can span distances in space measured in tens of thousands of miles. The possible applications of this type of Buck Rogers technology is only now beginning to be understood.

The most radical idea is Arthur Clarke's "Space Elevator" concept. The basic plan is to lower a tethered cable from geosynchronous orbit down to earth and tether it and also play out enough cable above Clarke orbit to "cancel" the mass of the cable as it becomes in effect weightless about the geosynchronous orbit. It may seem a little bit like black magic, but trust me the physics and the orbital mechanics do indeed work out based upon orbital speed and the pull of gravity.

Even Clarke concedes however, that such a tethered cable would be too costly. Further, the irregularities of the earth's size and density would snap the cable. If one could indeed build the space elevator, however, it would allow you to deploy satellites into low, medium or geosynchronous orbit at a small fraction of today's cost.

It is my view that if something can't work at the macro-scale, then perhaps a larger number of small steps could get you to a similar result. In this case there is the Pelton "Space Escalator" concept that suggests that if you take a large number of satellites (about 40 of them) each equipped with more modest length tethers of say 500 miles each and were to use ion engines to maneuver the satellite to be launched up from one "escalator step" to whatever level were desired within a matter of weeks.

With planning you could manage to deploy millions of tons of materials and equipment by virtually continuously "lifting" of payloads into orbit rather than using rocket launches.

Finally a satellite tethered from geosynchronous orbit into lower orbit could become geostationary at lower altitudes and thus like SPIDERSAT can minimize satellite delay. This concept would require deployment of "negative mass" above the geosynchronous orbit as well, and would mean that the first step down the escalator from geosynchronous orbit would be a great big one. After that the steps would be much more modest.

12.5.3 New Types of Satellite Architecture

The first operational satellite, "Early Bird", was quite simple. It looked like a Maxwell House coffee can with a pipe sticking out the top. Nevertheless, it had an antenna systems, solar cells and a battery power system and fuel to keep it in position. The present day satellite has up to 1,000 times greater capacity but the basic concept is the same. In many cases it even still looks like a big blue coffee can.

It now seems as if we are, as the Twenty-first Century approaches, on the verge of radically new architecture. What will be the key aspects of this architecture? The major elements that we may see between 2000 and 2015 are the following: (a) Intersatellite links (close and long range); (b) Remotely powered satellites; (c) Satellite clusters (two to eight satellites locked into a tight orbital configuration; (d) On-board switching and regeneration perhaps with optical switches; (e) Large satellite platforms with very extended lifetime e.g. 25 to 50 years; and (f) Satellite design for in-orbit servicing (fuel and batteries) or retrofit and upgrade to allow 25 year or more lifetimes.

This is not to say that all of these will occur. Some options may in fact tend to exclude others. The real point is that a wealth of new options seem likely to explode in the Twenty-first Century and that the "satellites" that results will not necessarily resemble today's "birds" or "coffee cans". This also suggests that the launch systems for these advanced satellites could also be radically different.

12.6 CONCLUSION

A final word about the long range future. Technology, even if it is exciting, with a high level of sex appeal will still not set the future course. Services and more particularly service demand by the consumer is the key to the future Twenty-first Century just as it was in the Nineteenth and Twentieth Centuries as well. If there is a demand for high definition television channels, it will create a need to design an appropriate satellite to respond, not <u>vice versa</u>. The current market indications are that the satellite market shapers will be: (a) <u>Mobile Services</u> - Aeronautical, maritime, land mobile and disaster and emergency relief communications systems will be crucial services because of their continuing growth and need for much higher power to reach very low-cost and portable receivers; (b) <u>Broadcast Services</u> - A host of new video and audio services are under development: High Definition Television; Multiple rastered TV; 3-D TV; and Educational and health-related videoconferencing are all high impact services that will affect the future direction of satellites. (c) <u>ISDN</u> — ISDN is far more of a puzzle. Once described as the world's best kept trillion secret, the Integrated Services Digital Network (ISDN) is nothing more or nothing less that a blueprint for creating the Blobal Electronic Village using digital communications as the architect. If ISDN does ultimately get as a service-pulled application by the consumer, rather than a supplier-pushed fad then ISDN will emerge as the number one market shaper. So far, two telephone channels and a data channel in every home is the best ISDN has mustered. A better appeal than this such as video phone, group 4 fax or some new unsuspected service will be needed to make ISDN the "future of communications".

Regardless of how new services evolve and which communications markets are changed, this much is clear: Satellites together with Fiber Optic Cables, will both be a major part of the future. For satellites, the future is now — and tomorrow too!

GLOSSARY
COMMONLY USED WORDS IN THE SATELLITE COMMUNICATIONS INDUSTRY

A

ACTS - The NASA experimental satellite project which demonstrates the use of the Ka Band (30/20 GHz) services, on-board processing with multiple beams, electronic hopping antenna beams, and dynamic adjustment of power to cope with rain attenuation. This highly successful experimental network which stands for Advanced Communications Technology Satellite can support low to medium rate mobile services to high data rate HDTV fixed services.

Adaptive Delta Modulation (ADM) - Developed by Dolby Labs, this digital coding scheme is utilized by Australia's satellite system for audio distribution.

Amplitude Modulation (AM) - The baseband signal is caused to vary the amplitude or height of the carrier wave to create the desired information content.

Amplifier - A device used to boost the strength of an electronic signal.

Analog - A form of transmitting information characterized by continuously variable quantities, as opposed to digital transmission, which is characterized by discrete bits of information in numerical steps. An analogue signal is responsive to changes in light, sound, heat and pressure.

Analog-to-Digital Conversion (ADC) - Process of converting analog signals to a digital representation. DAC represents the reverse translation.

ANIK - The Canadian domestic satellite system that transmits Canadian Broadcasting Corporation's (CBC) network feeds throughout the country. This system also carries long distance voice and data services throughout Canada as well as some transborder service to the US. and Mexico.

Antenna - A device for transmitting and receiving radio waves. Depending on their use and operating frequency, antennas can take the form of a single piece of wire, a di-pole a grid such as a yagi array, a horn, a helix, a sophisticated parabolic-shaped dish, or a phase array of active electronic elements of virtually any flat or convoluted surface.

Aperture - A cross sectional area of the antenna which is exposed to the satellite signal.

Apogee - The point in an elliptical satellite orbit which is farthest from the surface of the earth. Geosynchronous satellites which maintain circular orbits around the earth are first launched into highly elliptical orbits with apogees of 22,237 miles. When the communication satellite reaches the appropriate apogee, a rocket motor is fired to place the satellite into its permanent circular orbit of 22,237 miles.

Apogee Kick Motor (AKM) - Rocket motor fired to circulate orbit and deploy satellite into geostationary orbit.

Apstar-Asia-Pacific Star - Name of the recently deployed Chinese satellite system which is carrying commercial video service in the region. The first launch of this satellite was placed in an: "illegal" position at 131 degrees west adjacent to a Japanese NStar and a RIMSAT satellite. It was subsequently moved to another RIMSAT registered location in return for a financial settlement.

Arabsat - This is the Arabsat Satellite Organization and its is headquartered in Riyadh, Saudi Arabia. It provided regional telecommunications services for the Middle East region via a two satellite network.

Artemis - An experimental communications satellite package being designed by the European Space Agency to test on-board processing and signal regeneration. As such it is similar to ACTS and ETS VI. This is only an experimental payload and as such it may fly with other experimental packages.

ASCII - American Standard Code for Information Interchange. Standard computer system text coding format.

AsiaSat - A satellite system covering the Asia mainland. It is a joint venture of the Chinese government, Cable and Wireless and the Hutchenson Trading company.

Asynchronous Communications - Stream of data routed through a network as generated, rather than in organized message blocks. Most personal computers send data in this format. (See ATM)

Asynchronous Transfer Mode (ATM) - This is the new form of super-fast packet switching. In the 21st Century ATM networks will operate at speeds in the Gigabits/second.

Attenuation - The loss in power of electromagnetic signals between transmission and reception points.

Attitude Control - The orientation of the satellite in relationship to the earth and the sun.

Aries-A low earth orbit satellite system filed in the US. that would provide global mobile voice and data services.

AT&T Telstar-The domestic satellite system designed and deployed by AT&T for domestic US. satellite service.

Audio - Relating to sound or its reproduction. This is frequently used in transmission or reception of sound, music or voice.

AUSTEL - Australian Telecommunications. The regulatory agency for telecommunications in Australia. (See OFTEL).

Audio Subcarrier - The carrier between 5 MHz and 8 MHz containing audio (or voice) information inside of a video carrier.

Automatic Frequency Control (AFC) - A circuit which automatically controls the frequency of a signal.

Automatic Gain Control (AGC) - A circuit which automatically controls the gain of an amplifier so that the output signal level is virtually constant for varying input signal levels.

Az-El Mount - Antenna mount that requires two separate adjustments to move from one satellite to another; (See - Azimuth and Elevation).

Azimuth - The angle of rotation (horizontal) that a ground based parabolic antenna must be rotated through to point to a specific satellite in a geosynchronous orbit. The azimuth angle for any particular satellite can be determined for any point on the surface of the earth given the latitude and longitude of that point. It is defined with respect to due north as a matter of easy convenience.

B

B-Mac - A method of transmitting and scrambling television signals. In such transmissions MAC (Multiplexed Analog Component) signals are time-multiplexed with a digital burst containing digitized sound, video synchronizing, authorization and information.

Backhaul - A terrestrial communications channel linking an earth station to a local switching network or population center.

Backoff - The process of reducing the input and output power levels of a traveling wave tube to obtain more linear operation. This is sometimes called the terrestrial link.

Band Pass Filter - An active or passive circuit which allows signals within the desired frequency band to pass through but impedes signals outside this pass band from getting through.

Bandwidth - A measure of spectrum (frequency) use or capacity. For instance, a voice transmission by telephone requires a bandwidth of about 3000 cycles per second (3KHz). A TV channel occupies a bandwidth of 6 million cycles per second (6 MHz) in terrestrial systems. In satellite based systems a larger bandwidth of I7.5 to 72 MHz is used to spread or "dither" the TV signal in order to prevent interference.

Baseband - The basic direct output signal in an intermediate frequency based obtained directly from a television camera, satellite television receiver, or video tape recorder. Baseband signals can be viewed only on studio monitors. To display the baseband signal on a conventional television set a "remodulator" is required to convert the baseband signal to one of the VHF or UHF television channels which the television set can be tuned to receive.

Baud - The rate of data transmission based on the number of signal elements or symbols transmitted per second. Today most digital signals are characterized in bits per second.

Beacon - Low-power carrier transmitted by a satellite which supplies the controlling engineers on the ground with a means of monitoring telemetry data, tracking the satellite, or conducting propagation experiments. This tracking beacon is usually a horn or omni antenna.

Beamwidth - The angle or conical shape of the beam the antenna projects. Large antennas have narrower beamwidths and can pinpoint satellites in space or dense traffic areas on the earth more precisely. Tighter beamwidths thus deliver higher levels of power and thus greater communications performance.

BPSK - Binary phase Shift Keying is a digital modulation technique in which the carrier phase can have one of two possible values, namely 0 degrees or 180 degrees.

Bird - Slang for a communications satellite located in geosynchronous orbit.

Bit - A single digital unit of information

Bit Error Rate - The fraction of a sequence of message bits that are in error. A bit error rate of 10-6 means that there is an average of one error per million bits.

Bit Rate - The speed of a digital transmission, measured in bits per second.

Blanking - An ordinary television signal consists of 30 separate still pictures or "frames" sent every second. They occur so rapidly, the human eye blurs them together to form an illusion of moving pictures. This is the basis for television and motion picture systems. The blanking interval is that portion of the television signal which occurs after one picture frame is sent and before the next one is transmitted. During this period of time special data signals can be sent which will not be picked up on an ordinary television receiver.

Block Down converter (LNB) - A device used to convert the 3.7 to 4.2 GHz signal down to UHF or lower frequencies (1 GHz and lower).

BNC Connector - A standard medium-sized twistlock coaxial cable connector used at higher frequencies throughout the television industry and by many satellite TV receivers. Other popular connectors include the larger screw in Type N and the smaller miniature-type connectors.

Boresight- Axis of symmetry of a paraboloid.

Broadcasting - The process of transmitting a radio or television signal via an antenna to multiple receivers which can simultaneously pick up the signal. This can be accomplished from space or by means of ground antenna systems. Cable Television has become one of the most important transmission media in competition with traditional broadcasting.

BSS - Broadcast Satellite Service - This is the ITU designation but DBS or Direct Broadcast Service is more commonly used term in the satellite industry.

Business Television - Corporate communications tool involving video transmissions of information via satellite. Common uses of business television are for meetings, product introductions and training.

Buttonhook Feed - A shaped piece of waveguide directing signal from the feed to the LNA behind the antenna.

Bypass - Use of satellite, local area network, wide area network, or metropolitan area network as an alternative transmission facility. Most frequently this is for data and the usual purpose is to avoid the local telephone company network in order to achieve low cost services.

C

C-Band - This is the band between 4 and 8 GHz with the 6 and 4 GHz band being used for satellite communications. Specifically, the 3.7 to 4.2 GHz satellite communication band is used as the downlink frequencies in tandem with the 5.925 to 6,425 Ghz band that serves as the uplink.

Carrier to Noise Ratio (C/N) - The ratio of the received carrier power and the noise power in a given bandwidth, expressed in db. This figure is directly related to G/T and S/N; and in a video signal the higher the C/N, the better the received picture.

Carrier - The basic radio, television, or telephony center of frequency transmit signal. The carrier in an analog signal is modulated by manipulating its amplitude (making it louder or softer) or its frequency (shifting it up or down) in relation to the incoming signal. Satellite carriers operating in the analog mode are usually frequency modulated.

Carrier Frequency - The main frequency on which a voice, data, or video signal is sent. Microwave and satellite communications transmitters operate in the band from 1 to 14 gigahertz (a gigahertz is one billion cycles per second).

Cassegrain Antenna - The antenna principle that utilizes a subreflector at the focal point which reflects energy to or from a feed located at the apex of the main reflector.

CATV - Originally meant Community Antenna Television. Independent "mom and pop" companies in rural communities would build a large television receiving antenna on a nearby mountain to pick up the weak TV signals from a distant metropolis. These signals were amplified, modulated onto television channels and sent along a coaxial cable strung from house to house. Now standing for Cable Television; most independent CATV companies have long since been purchased by national organizations that own multiple cable systems in rural and urban areas.

CCITT - (now TSS) Comite Consultatif Internationale de Telegraphique et Telephonique. International body, associated with the ITU, which establishes worldwide standards for telecommunications. Reorganized to include CCIR (radio standards group) and renamed TSS (Telecommunications Standardization Sector).

CDMA - Code division multiple access. Refers to a multiple-access scheme where stations use spread-spectrum modulations and orthogonal codes to avoid interfering with one another. See Spread Spectrum.

Channel - A frequency band in which a specific broadcast signal is transmitted. Channel frequencies are specified in the United States by the Federal Communications Commission. Television signals require a 6 megahertz frequency band to carry all the necessary picture detail.

CIF - Common Intermediate Format - a compromise television display format adopted by the CCITT which is relatively easy to derive from both PAL and NTSC.

Circular Polarization - Unlike many domestic satellites which utilize vertical or horizontal polarization, the international Intelsat satellites transmit their signals in a rotating corkscrew-like pattern as they are down-linked to earth. On some satellites, both right-hand rotating and left-hand rotating signals can be transmitted simultaneously on the same frequency; thereby doubling the capacity of the satellite to carry communications channels.

Clamp - A video processing circuit that removes the energy dispersal signal component from the video waveform.

Clarke Orbit - That circular orbit in space 22,237 miles from the surface of the earth at which geosynchronous satellites are placed. This orbit was first postulated by the science fiction writer Arthur C. Clarke in Wireless World magazine in 1945. Satellites placed in these orbits, although traveling around the earth at thousands of miles an hour, appear to be stationary when viewed from a point on the earth, since the earth is rotating upon its axis at the same angular rate that the satellite is traveling around the earth.

C/No or C/kTB - Carrier-to-noise ratio measured either at the Radio Frequency (RF) or Intermediate Frequency (IF)

Coaxial Cable - A transmission line in which an inner conductor is surrounded by an outer conductor or shield and separated by a nonconductive dielectric, typically a foam. Coax cables have the capacity to carry enormously high frequency signals in the television range, and thus are used by CATV companies for the signal distribution. A number of cable TV systems are moving over to fiber optic cable as their backbone distribution system.

Codec - Coder/Decoder, a data processing device which converts analog signals to digital form and vice versa. Codecs also compress and decompress data.

Co-Location - Ability of multiple satellites to share the same approximate geostationary orbital assignment frequently due to the fact that different frequency bands are used.

Color Subcarrier - A subcarrier that is added to the main video signal to convey the color information. In NTSC systems, the color subcarrier is centered on a frequency of 3.579545 MHz, referenced to the main video carrier.

COMETS - A Japanese experimental satellite program to test new concepts and frequencies for mobile and DBS satellite services. Two satellites are projected in this series.

Common Carrier - Any organization which operates communications circuits used by other people. Common carriers include the telephone companies as well as the owners of the communications satellites, RCA, Western Union, Comsat General, AT&T and others. Common carriers are required to file fixed tariffs for specific services.

Companding - A noise-reduction technique that applies single compression at the transmitter and complementary expansion at the receiver.

Composite Baseband - The unclamped and unfiltered output of the satellite receiver's demodulator circuit, containing the video information as well as all transmitted subcarriers.

Compression Algorithms - Software that allows codecs to reduce the number of bits required for data storage or transmission.

COMSAT - The Communications Satellite Corporation which serves as the US. Signatory to INTELSAT and INMARSAT.

Conus - Contiguous United States. In short, all the states in the US. except Hawaii and Alaska.

CPE - Customer Premises Equipment such as telephones, terminals, codecs, multiplexers, etc.

Cross Modulation - A form of signal distortion in which modulation from one or more RF carrier(s) is imposed on another carrier.

CSU- Channel Service Unit. A CPE component which terminates a digital circuit such as T1. CSU assures compliance to FCC regulations and performs some line-conditioning functions.

C/T - Carrier-to-noise-temperature ratio.

D

DAMA - Demand-Assigned Multiple Access - A highly efficient means of instantaneously assigning telephony channels in a transponder according to immediate traffic demands.

DBS - Direct broadcast satellite. Refers to service that uses satellites to broadcast multiple channels of television programming directly to home mounted small-dish antennas.

dBW - The ratio of the power to one Watt expressed in decibels.

De-BPSK - Differential Binary Phase Shift Keying

De-QPSK- Differential Quadrature Phase Shift Keying.

DCE - Data Circuit-Terminating Equipment. Equipment at a node or access point of a network.

Decibel (db) - The standard unit used to express the ratio of two power levels. It is used in communications to express either a gain or loss in power between the input and output devices.

Declination - The offset angle of an antenna from the axis of its polar mount as measured in the meridian plane between the equatorial plane and the antenna main beam.

Decoder - A television set-top device which enables the home subscriber to convert an electronically scrambled television picture into a viewable signal. This should not be confused with a digital coder/ decoder known as a CODEC which is used in conjunction with digital transmissions.

Deemphasis - Reinstatement of a uniform baseband frequency response following demodulation.

Delay - The time it takes for a signal to go from the sending station through the satellite to the receiving station. This transmission delay for a single hop satellite connection is very close on one-quarter of a second.

Demodulator - A satellite receiver circuit which extracts or "demodulates" the "wanted" signals from the received carrier.

Deviation - The modulation level of an FM signal determined by the amount of frequency shift from the frequency of the main carrier.

Differential Gain/Phase - Nonlinear color video distortion parameters.

Digital - Conversion of information into bits of data for transmission through wire, fiber optic cable, satellite, or over air techniques. Method allows simultaneous transmission of voice, data or video.

Digital Speech Interpolation - DSI - A means of transmitting telephony. Two and one half to three times more efficiently based on the principle that speakers are transmitting only about 40% of the time.

Discriminator - A type of FM demodulator used in satellite receivers.

Distribution Amplifier - Wide-band amplifier operating at the VHF television channel frequencies used by the CATV companies to periodically strengthen the weakened signals as they are transmitted down by the CATV company's cable network. Distribution amplifiers are also used in apartment and other master antenna television (MATV) installations. Some models operate on baseband signals as well.

Distribution Center - The central point from which the television signals are distributed from a television programming organization to its network stations or receivers. The distribution center for a satellite television programmer will usually consist of a bank of video tape machines, studio facilities, and large uplink (transmit) antenna for transmission of the signal directly to the desired satellite.

Dithering - The process of shifting the 6-MHz satellite-TV signal up and down the 36-MHz satellite transponder spectrum at a rate of 30 times per second (30 Hertz). The satellite signal is "dithered" to spread the transmission energy out over a band of frequencies far wider than a terrestrial common carrier microwave circuit operates within, thereby minimizing the potential interference that any one single terrestrial microwave transmitter could possibly cause to the satellite transmission.

Down-Converter - That portion of the Fixed Satellite Service (FSS) television receiver that converts the signals from the 4-GHz microwave range to (typically) the more readily used baseband or intermediate frequency (IF)70-MHz range. Down-converters typically were located physically near the receiver, requiring bulky and expensive coaxial cable feeds from the antenna to the receiver. Newer designs have seen the down-converter placed at the antenna itself often in combination with the Low Noise Amplifier (LNA) thus allowing miniature CATV like coaxial cable to bring the satellite TV signal into the house. This type of design allows more flexibility in the location of the antenna with respect to the house. Block down-converters are typically used today.

Downlink - The 4 GHz frequency range utilized by the satellite to retransmit signals down to earth for reception.

DSU - Data Service Unit. An equipment component that resides at the customer premise, which interfaces to a digital circuit such as T1 or Switched 56. DSU performs conversion of data stream to bipolar format for transmission. Generally used or combined with a CSU.

Dual Spin - Spacecraft design whereby the main body of the satellite is spun to provide altitude stabilization, and the antenna assembly is despun by means of a motor and bearing system in order to continually direct the antenna earthward. This dual-spin configuration thus serves to create a spin stabilized satellite.

E

Earth Station - The term used to describe the combination or antenna, low-noise amplifier (LNA), down-converter, and receiver electronics used to receive a signal transmitted by a satellite. Earth Station antennas vary in size from the 2 foot to 10 foot (65 centimeters to 3 meters) diameter size used for TV reception to as large as 100 feet (30 meters) in diameter sometimes used for international communications. The typical antenna used for INTELSAT communication is today I3 to I8 meters or 40 to 60 feet.

Echo Canceller - An electronic circuit which attenuates or eliminates the echo effect on satellite telephony links. Echo cancellers are largely replacing obsolete echo suppressors.

Echo Effect - A time-delayed electronic reflection of a speaker's voice. This is largely eliminated by modern digital echo cancellers (see above).

Eclipse - When a satellite passes through the line between the earth and the sun or the earth and the moon.

Eclipse Protected - Refers to a transponder that can remain powered during the period of an eclipse.

Edge of Coverage - Limit of a satellite's defined service area. In many cases, the EOC is defined as being 3 dB down from the signal level at beam center. However, reception may still be possible beyond the 3-dB point.

EEO - Extremely Elliptical Orbit (See HEO).

EIRP (Effective Isotropic Radiated Power) - This term describes the strength of the signal leaving the satellite antenna or the transmitting earth station antenna, and is used in determining the C/N and S/N. The transmit power value in units of dBW is expressed by the product of the transponder output power and the gain of the satellite transmit antenna.

El/Az- An antenna mount providing independent adjustments in elevation and azimuth.

Elevation - The upward tilt to a satellite antenna measured in degrees required to aim the antenna at the communications satellite. When aimed at the horizon, the elevation angle is zero. If it were tilted to a point directly overhead, the satellite antenna would have an elevation of 90 degrees.

Ellipso - A system filed in the US. that would provide mobile voice and data services by a combination of leo satellites and elliptical orbit satellites so as to better serve higher latitude locations such as the US.

Encoder (scrambler) - A device used to electronically alter a signal so that it can only be viewed on a receiver equipped with a special decoder.

Energy Dispersal - A low-frequency waveform combined with the baseband signal prior to modulation, to spread the FM signal's peak power across the available transponder bandwidth in order to reduce the potential for creating interference to ground-based communications services.

EOL - End of Life of a satellite.

Equatorial Orbit - An orbit with a plane parallel to the earth's equator.

ESC - Engineering Service Circuit - The 300-3,400 Hertz voice plus teletype (S+DX) channel used for earth station-to-earth station and earth station-to-operations center communications for the purpose of system maintenance, coordination and general system information dissemination. In analog (FDM/FM) systems there are two S+DX channels available for this purpose in the 4,000-12,000 Hertz portion of the baseband. In digital systems there are one or two channels available which are usually converted to a 32 or 64 Kb/s digital signal and combined with the earth station traffic digital bit stream. Modern ESC equipment interfaces with any mix of analog and digital satellite carriers, as well as backhaul terrestrial links to the local switching center.

ETS - Experimental Test Satellite. This is an on-going experimental communications program of NASDA and the Ministry of

Post and Telecommunications of Japan. There have been six satellites in this series to date. The most advanced of these satellites, the ETS VI , with many features similar to the US. ACTS satellite was launched into an elliptical medium earth orbit which limits its effectiveness. This satellite will also test optical communications links as well. The ETS VII will experiment with robotic self-repair in orbit.

Eutelsat - The European Telecommunications Satellite Organization which is headquartered in Paris, France. It provides a satellite network for West and Eastern Europe and parts of North Africa and the Middle East. Its primary service is for video distribution but it also provides VSAT business services known as SMS.

F

FCC (Federal Communications Commission) - The US. federal regulatory body, consisting of five members, one of who is designated chairman, appointed by the President and confirmed by the Senate, which regulates interstate communications under the Communications Act of 1934.

F/D - Ratio of antenna focal length to antenna diameter. A higher ratio means a shallowed dish.

FDMA - Frequency division multiple access. Refers to the use of multiple carriers within the same transponder where each uplink has been assigned frequency slot and bandwidth. This is usually employed in conjunction with Frequency Modulation.

Feed - This term has at least two key meanings within the field of satellite communications. It is used to describe the transmission of video programming from a distribution center. It is also used to describe the feed system of an antenna. The feed system may consist of a subreflector plus a feedhorn or a feedhorn only.

Feedhorn - A satellite TV receiving antenna component that collects the signal reflected from the main surface reflector and channels this signal into the low-noise amplifier (LNA)

FM - Frequency Modulation - A modulation method whereby the baseband signal varies the frequency of the carrier wave.

FM Threshold - That point at which the input signal power is just strong enough to enable the receiver demodulator circuitry successfully to detect and recover a good quality television picture from the incoming video carrier. Using threshold extension techniques, a typical satellite TV receiver will successfully provide good pictures with an incoming carrier noise ratio of 7db. Below the threshold a type of random noise called "sparkles" begins to appear in the video picture. In a digital transmission, however, signal is suddenly and dramatically lost when performance drops under the threshold.

Focal Length - Distance from the center feed to the center of the dish.

Focal Point - The area toward which the primary reflector directs and concentrates the signal received.

Footprint - A map of the signal strength showing the EIRP contours of equal signal strengths as they cover the earth's surface. Different satellite transponders on the same satellite will often have different footprints of the signal strength. The accuracy of EIRP footprints or contour data can improve with the operational age of the satellite. The actual EIRP levels of the satellite, however, tends to decrease slowly as the spacecraft ages.

Forward Error Correction (FEC) - Adds unique codes to the digital signal at the source so errors can be detected and corrected at the receiver. These codes can include linear block codes, binary cyclic codes, and convolutional codes.

Frequency - The number of times that an alternating current goes through its complete cycle in one second of time. One cycle per second is also referred to as one hertz; 1000 cycles per second, one kilohertz; 1,000,000 cycles per second, one megahertz; and 1,000,000,000 cycles per second, one gigahertz.

Frequency-Agile - The ability of a satellite TV receiver to select or tune all 12 or 24 channels (transponders) from a satellite. Receivers not frequency-agile are dedicated to a single channel and are most often used in the CATV industry. Frequency agility can be via continuously variable tuning or discreet step (channel selection) tuning.

Frequency Coordination - A process to eliminate frequency interference between different satellite systems or between terrestrial microwave systems and satellites.. In the US. this activity relies upon a computerized service utilizing an extensive database to analyze potential microwave interference problems that arise between organizations using the same microwave band. As the same C-band frequency spectrum is used by telephone networks and CATV companies when they are contemplating the installation of an earth station, they will often obtain a frequency coordination study to determine if any problems will exist.

Frequency Reuse - A technique which maximizes the capacity of a communications satellite through the use of spacially isolated beam antennas and/or the use of dual polarities. Up to six-fold frequency reuse has been achieved in the C-Band on the INTELSAT VI satellite series.

G

Gain - A measure of amplification expressed in db. (See detailed explanation provided in the text).

GE Americon - This is a large US. satellite system for domestic communications. Since it consolidates the former RCA system, the GTE system and the pre—existing GE systems, it is now the largely domestic US. satellite operation

Geostationary - Refers to a geosynchronous satellite angle with zero inclination, so the satellite appears to hover over one spot on the earth's equator.

Geostationary Transfer Orbit - This orbit is in the equatorial plane. This type of orbit has an elliptical form, with a perigee at 200 km and an apogee at 35870 km. Ariane injects geostationary satellites into orbits of this type.

Geosynchronous - The Clarke circular orbit above the equator. For a planet the size and mass of the earth, this point is 22,237 miles above the surface.

Gigahertz (GHz) - One billion cycles per second. Signals operating above 3 Gigahertz are known as microwaves. Above 30 GHz they are know as millimeter waves. As one moves above the millimeter waves signals begin to take on the characteristics of lightwaves.

Global Beam - An antenna down-link pattern used by the Intelsat satellites, which effectively covers one-third of the globe. Global beams are aimed at the center of the Atlantic, Pacific and Indian Oceans by the respective Intelsat satellites, enabling all nations on each side of the ocean to receive the signal. Because they transmit to such a wide area, global beam transponders have significantly lower EIRP outputs at the surface of the Earth as compared to a US domestic satellite system which covers just the continental United States. Therefore, earth stations receiving global beam signals need antennas much larger in size (typically 10 meters and above (i.e. 30 feet and up).

Globalstar - A proposed mobile satellite system that would deploy a network of 48 satellites to create a global voice and data service. This system is backed by Qualcomm, Loral, and Alcatel. It lacks the technical complexity of the Iridium system in that it is a "bent-pipe" system without intersatellite links.

Gonets - A proposed Russian satellite system that would be a so-called "little leo" system to provide store and forward data services on a global basis. The frequency proposed for Gonets is not approved for US. usage since it interferes with a US. military frequency allocation.

Gregorian - Dual-reflector antenna system employing a paraboloidal main reflector and a concave ellipsoidal subreflector.

G/T - A figure of merit of an antenna and low noise amplifier combination expressed in db. "G" is the net gain of the system and "T" is the noise temperature of the system. The higher the number, the better the system.

Guard Channel - Television channels are separated in the frequency spectrum by spacing them several megahertz apart. This unused space serves to prevent the adjacent television channels from interfering with each other.

H

Half-Transponder - A method of transmitting two TV signals through a single transponder through the reduction of each TV signal's deviation and power level.

Headend - Electronic control center - generally located at the antenna site of a CATV system - usually including antennas, preamplifiers, frequency converters, demodulators and other related equipment which amplify, filter and convert incoming broadcast TV signals to cable system channels.

Heliosynchronous Orbit - (altitude 600 to 800 km.) Situated in a quasi-polar plane. The satellite is permanently visible from that part of the Earth in sunlight. Heliosynchronous orbits are used for Earth observation or solar-study satellites.

HEO - Highly Elliptical Orbit. This is type of orbit used by the Russian Molniya Satellite system. It is also referred to as Extremely Elliptical Orbit (EEO).

Hertz (Hz) - The name given to the basic measure of radio frequency characteristics. An electromagnetic wave completes a full oscillation from its positive to its negative pole and back again in what is known as a cycle. A single Hertz is thus equal to one cycle per second.

High Band - TV channels 7 through 13.

High Frequency (HF) - Radio frequencies within the range of 3,000 to 30,000 kilohertz. HF radio is known as shortwave.

High-Power Satellite - Satellite with 100 watts or more of transponder RF power.

Hour Angle - Steering direction of a polar mount. An angle measured in the equatorial plane between the antenna beam and the meridian plane.

Hub - The master station through which all communications to, from and between micro terminals must flow. In the future satellites with on-board processing will allow hubs to be eliminated as MESH networks are able to connect all points in a network together.

Hughes Galaxy - A domestic US. satellite system which provides a range of telecommunications services but is primarily a video distribution network for networks and cable television systems.

Hughes Spaceway - A proposed very high powered geosynchronous satellite system that would utilized the Ka Band (30/20 GHz) and operate to USAT microterminals.

I

IBS - INTELSAT Business Service

IFRB - International Frequency Registration Board of the ITU

Inclination - The angle between the orbital plane of a satellite and the equatorial plane of the earth.

INMARSAT - The International Maritime Satellite Organization operates a network of satellites for international transmissions for all types of international mobile services including maritime, aeronautical, and land mobile.

INMARSAT P - The proposed INMARSAT satellite system for personal mobile communications that would involve a network of medium earth orbit satellites that could connect to the INMARSAT geosynchronous satellite system. This project as now conceived would be a privatized and fully competitive system separate for the regular network which was designed for maritime and aeronautical services.

INTELSAT - The International Telecommunications Satellite Organization operates a network of 20 satellites primarily for international transmissions but which provides domestic services to some 40 countries as well. Current membership is over 130 countries and satellite services are provided to nearly 200 countries and territories.

Interactive Cable - Systems having capabilities to send signals upstream and downstream (see two-way capacity).

Interface - Common boundary between two or more items of equipment or between a terminal and a communication channel. Also, the electronic device to interconnect two or more devices or items of equipment having similar or dissimilar characteristics. The electronic device placed between a terminal and a communication channel to protect the network from the hazard of excess voltage levels.

Interference - Energy which tends to interfere with the reception of the desired signals, such as fading from airline flights, RF interference from adjacent channels, or "hosting from reflecting objects such as mountains and buildings.

Inter Satellite Link - Inter Satellite Links or ISLs are radio or optical communications links between satellites which may be co-located, within the same region, or inter-regional in the case of geosynchronous satellite systems. They can also connect low earth orbit and GEO satellites as in the case of the NASA Tracking and Data Relay Satellite. Finally they can serve to interconnect global constellations of satellites in low and medium earth orbit as well.

INTERSPUTNIK - The international entity formed by the Soviet Union to provide international communications via a network of Soviet satellites. Its primary travel is video distribution and its primary users are former members of the Soviet Union and approximately a dozen other countries such as Cuba, Burma, Algeria, Vietnam, etc.

Iridium Satellite System - This is a 66 satellite network proposed by the Motorola corporation in a US. filing for global mobile voice and data services. It is planned as a global consortium with multiple owners around the world. Each satellite would operate with 48 cells to allow very intensive re-use of frequencies. This system unlike most of the others which propose to use CDMA multiplexing plans to use TDMA multiplexing instead.

ISDN - Integrated Services Digital Network. A CCITT standard for integrated transmission of voice, video and data. Bandwidths include: Basic Rater Interface - BRI (144 Kbps) and Primary Rate - PRI (1.544 and 2.048 Mbps).

ISO - International Standards Organization. Develops standards such as JPEG and MPEG. Closely allied with the CCITT.

Isotropic Antenna - A hypothetical omnidirectional point-source antenna that serves as an engineering reference for the measurement of antenna gain.

ITALSAT - An experimental satellite designed by the Italian government to test satellite communications and especially rain fade in the millimeter wave region.

ITU - International Telecommunication Union.

J

JPEG - ISO Joint Picture Expert Group standard for the compression of still pictures.

K

Ka-Band - The frequency range from 18 to 31 GHz. The spectrum allocated for satellite communication is 30 GHz for the up-link and 20 GHz for the down-link.

Kb/s - Kilobits per second. Refers to transmission speed of 1,000 bits per second.

Kelvin (K) - The temperature measurement scale used in the scientific community. Zero K represents absolute zero, and corresponds to minus 459 degrees Fahrenheit or minus 273 Celsius. Thermal noise characteristics of an LNA are measured in Kelvins. NOTE. Unlike other temperature measurements, the word "degree" (or symbol) is not used when expressing temperatures in Kelvins.

Kilohertz (kHz) - Refers to a unit of frequency equal to 1,000 Hertz.

Klystron - A type of high-power amplifier which uses a special beam tube.

Ku-Band - The frequency range from 10.9 to 17 GHz. The spectrum used for satellite communications is 14 GHz for uplink and 12/11 GHz for the downlink.

L

L-Band - The frequency range from 0.5 to 1.5 GHz. Also used to refer to the 950- to 1450MHz used for mobile communications.

Leased Line - A dedicated circuit typically supplied by the telephone company.

Low-Band - TV channels 2 through 6.

Low Noise Amplifier (LNA) - This is the preamplifier between the antenna and the earth station receiver. For maximum effectiveness, it must be located as near the antenna as possible, and is usually attached directly to the antenna receive port. The LNA is especially designed to contribute the least amount of thermal noise to the received signal.

Low Noise Converter (LNC) - A combination Low Noise Amplifier and down converter built into one antenna-mounted package.

Low Orbit - (altitude of 200 to 300 km.) This is used for certain types of scientific or observation satellites, which can view a different part of the Earth beneath them on each orbit revolution, as they overfly both hemispheres. The American Space Shuttles are injected into low-altitude orbits.

Low-Power Satellite - Satellite with transmit RF power below 30 watts.

Low Power TV (LPTV) - A new television service established by the Federal Communications Commission in October of 1980. LPTV broadcasting stations typically radiate between 100 and 1000 watts or power, covering a geographic radius of 10 to 15 miles. Upwards of 5000 LPTV stations will ultimately be licensed for operation. Many of these stations will obtain their programming from new satellite television networks now being formed.

M

MAC (A-, B-, C-, D2-) - Multiplexed analog component color video transmission system. Subtypes refer to the various methods used to transmit audio and data signals.

Margin - The amount of signal in dB by which the satellite system exceeds the minimum levels required for operation.

Master Antenna Television (MATV) - An antenna system that serves a concentration of television sets such as in apartment buildings, hotels or motels.

Medium-Power Satellite - Satellite generating transmit power levels ranging from 30 to 100 watts.

Megahertz (MHz) - Refers to a frequency equal to one million Hertz, or cycles per second.

Microwave - Line-of sight, point-to-point transmission of signals at high frequency. Many CATV systems receive some television signals from a distant antenna location with the antenna and the system connected by microwave relay. Microwaves are also used for data, voice, and indeed all types of information transmission. The growth of fiber optic networks have tended to curtail the growth and use of microwave relays.

Microwave Interference - Interference which occurs when an earth station aimed at a distant satellite picks up a second, often stronger signal, from a local telephone terrestrial microwave relay transmitter. Microwave interference can also be produced by nearby radar transmitters as well as the sun itself. Relocating the antenna by only several feet will often completely eliminate the microwave interference.

Mid Band - The part of the frequency band that lies between television channels 6 and 7, reserved by the FCC for air, maritime and land mobile units, FM radio and aeronautical and maritime navigation. Mid band frequencies 108 to 174 MHz can also be used to provide additional channels on cable television systems.

Modem - A communications device that modulates signals at the transmitting end and demodulates them at the receiving end.

Modulation - The process of manipulating the frequency or amplitude of a carrier in relation to an incoming video, voice or data signal.

Modulator - A device which modulates a carrier. Modulators are found as components in broadcasting transmitters and in satellite transponders. Modulators are also used by CATV companies to place a baseband video television signal onto a desired VHF or UHF channel. Home video tape recorders also have built-in modulators which enable the recorded video information to be played back using a television receiver tuned to VHF channel 3 or 4.

Molniya - The Russian domestic satellite system which operated with three highly elliptical satellites which took advantage of the inclination of these satellites which overlooked the high latitudes of the territories of the USSR.

MPEG-1 - ISO Motion Pictures Experts Group standard for the compression of motion or still video for transmission or storage.

MPEG-2 - This is the name given to the new International video compression standard.

Multiple Access - The ability of more than one user to have access to a transponder.

Multiple System Operator (MSO) - A company that operates more than one cable television system.

Multiplexing - Techniques that allow a number of simultaneous transmissions over a single circuit. TDMA and CDMA are the most common modes in satellite communications.

Multipoint Distribution System (MDS) - A common carrier licensed by the FCC to operate a broadcast-like omni directional microwave transmission facility within a given city. MDS carriers often pick up satellite pay-tv programming and distribute it via their local MDS transmitter to specially installed antennas and receivers in hotels, apartment buildings, and individual dwellings throughout the area.

Mux - A Multiplexer. Combines several different signals (e.g. video, audio, data) onto a single communication channel for transmission. Demultiplexing separates each signal at the receiving end.

N

NAB - National Association of Broadcasters.

NASA - National Aeronautics and Space Administration. The US. agency which administers the American space program, including the deployment of commercial and military satellites via a fleet of space shuttle vehicles.

NASDA - National Space Development Agency of Japan.

NCTA - National Cable Television Association.

NHK - Japanese broadcasting network distributed via BS-2 DBS satellites.

NIST - National Institute for Standards and Technology, a unit of the US. Department of Commerce that establishes standards in many areas including electronics and telecommunications.

Noise - Any unwanted and unmodulated energy that is always present to some extent within any signal.

Noise Figure (NF) - A term which is a figure of merit of a device, such as an LNA or receiver, expressed in db, which compares the device with a perfect device.

NTSC - National Television Standards Committee. A video standard established by the United States (RCA/NBC) and adopted by numerous other countries. This is a 525-line video with 3.58MHz chroma subcarrier and 60 cycles per second.

NTIA - National Telecommunications and Information Administration, a unit of the Department of Commerce that address US. government telecommunications policy, standards setting, and radio spectrum allocation.

Nutation Damping - The process of correcting the nutational effects of a spinning satellite which are similar in effect to a wobbling top. Active nutation controls use thruster jets.

O

Odyssey - A proposed medium earth orbit satellite system for voice and data services that would provide global service with 12 satellites in 10, 350 kilometer high circular orbits inclined 55 degrees from the equatorial plane.

OFTEL - The Office of Telecommunications of the United Kingdom government. This unit a part of the Department of Industries regulates telecommunications in the United Kingdom largely through maintaining open competition and fining entities who violate governmental guidelines for fair market practices.

Olympus - An experimental satellite program of the European Space Agency that demonstrated the use of Ka-Band (30/20 GHz) frequencies for space broadcasting.

Orbita - The name of the Russian domestic television earth station network which includes some 50 sites in the Russian territory.

Orbital Period - The time that it takes a satellite to complete one circumnavigation of its orbit.

OrbCom-The US. licensed 48 satellite "little leo" system for global store and forward services. This system is wholly owned by the Orbital Sciences Corporation (OSC) and will be deployed on the Pegasus launch vehicles also developed by OSC.

P

Packet Switching - Data transmission method that divides messages into standard-sized packets for greater efficiency of routing and transport through a network.

PAL - See Phase Alternation System. The German developed TV standard based upon 50 cycles per second and 625 lines.

PAM (PAM-D) - Payload assist module. A type of PKM.

PBS - Public Broadcasting System (US).

Parabolic Antenna - The most frequently found satellite TV antenna, it takes its name from the shape of the dish described mathematically as a parabola. The function of the parabolic shape is to focus the weak microwave signal hitting the surface of the dish into a single focal point in front of the dish. It is at this point that the feedhorn is usually located.

Pay-Per-View - A system of television in which scrambled signals are distributed and are unscrambled at the homeowner's set with a decoder that responds upon payment of a fee for each program. Pay TV can also refer to a system when subscribers pay an extra fee for access to a special channel, which might offer sports programs, first-run movies or professional training.

Perigee - The point in an elliptical satellite orbit which is closest to the surface of the earth.

Perigee Kick Motor (PKM) - Rocket motor fired to inject a satellite into a geostationary transfer orbit from a low earth orbit especially that of a STS or Shuttle-based orbit of 300 to 500 miles altitude.

Period - The amount of time that a satellite takes to complete one revolution of its orbit.

Phase Alternation System (PAL) - A European color television system incompatible with the US television system. US and Canadian satellites utilize the NTSC color television transmission system. Intelsat satellites often use the PAL system making them incompatible with US receivers.

Phase-Locked Loop (PLL) - A type of electronic circuit used to demodulate satellite signals.

Polarization - A technique used by the satellite designer to increase the capacity of the satellite transmission channels by reusing the satellite transponder frequencies. In linear cross polarization schemes, half of the transponders beam their signals to earth in a vertically polarized mode; the other half horizontally polarize their down links. Although the two sets of frequencies overlap, they are 90 degree out of phase, and will not interfere with each other. To successfully receive and decode these signals on earth, the earth station must be outfitted with a properly polarized feedhorn to select the vertically or horizontally polarized signals as desired. In some installations, the feedhorn has the capability of receiving the vertical and horizontal transponder signals simultaneously, and routing them into separate LNAs for delivery to two or more satellite television receivers. Unlike most domestic satellites, the Intelsat series use a technique known as left-hand and right-hand circular polarization.

Polarization Rotator - A device that can be manually or automatically adjusted to select one of two orthogonal polarizations.

Polar Mount - Antenna mechanism permitting steering in both elevation and azimuth through rotation about a single axis. While an astronomer's polar mount has its axis parallel to that of the earth, satellite earth stations utilize a modified polar mount geometry that incorporates a declination offset.

Polar Orbit - An orbit with its plane aligned in parallel with the polar axis of the earth.

Processor - An electronic or optical device that can "process" data usually at very high speed.

Protected-Use Transponder - A satellite transponder provided by the common carrier to a programmer with a built-in insurance policy. If the protected-use transponder fails, the common carrier guarantees the programmer that it will switch over to another transponder, sometimes pre-empting some other non-protected programmer from the other transponder.

PTT - Post Telephone and Telegraph Administration. Refers to operating agencies directly or indirectly controlled by governments in charge of telecommunications services in most countries of the world.

Pulse Code Modulation - A time division modulation technique in which analog signals are sampled and quantized at periodic intervals into digital signals. The values observed are typically represented by a coded arrangement of 8 bits of which one may be for parity.

Q

QPSK - Quadrature Phase Shift Keying is a digital modulation technique in which the carrier phase can have one of four possible value of 0, 90, 180, 270 degrees on the equivalent of a 90 degree rotation. There are even more advanced concepts based upon 8-phase (45 degree rotation), 16 phase (22.5 degree rotation) and so on to 32 phase, etc.

R

Rain Outage - Loss of signal at Ku or Ka -Band frequencies due to absorption and increased sky-noise temperature caused by heavy rainfall.

Receiver (Rx) - An electronic device which enables a particular satellite signal to be separated from all others being received by an earth station, and converts the signal format into a format for video, voice or data.

Receiver Isolation - The "isolation" between any two receivers connected to the system.

Receiver Sensitivity - Expressed in dBm this tells how much power the detector must receive to achieve a specific base band performance, such as a specified bit-error rate or signal to noise ratio.

RimSat - A satellite system which is US. owned but which is making use of under utilized Russian satellites by redeploying satellites originally filed for by the Tongasat Organization.

RF Adaptor - An add-on modulator which interconnects the output of the satellite television receiver to the input (antenna terminals) of the user's television set. The RF adaptor converts the baseband video signal coming from the satellite receiver to a radio frequency RF signal which can be tuned in by the television set on VHF channel 3 or 4.

S

Satellite - A sophisticated electronic communications relay station orbiting 22,237 miles above the equator moving in a fixed orbit at the same speed and direction of the earth (about 7,000 mph east to west).

Satellite Terminal - A receive-only satellite earth station consisting of an antenna reflector (typically parabolic in shape), a feedhorn, a low-noise amplifier (LNA), a down converter and a very sensitive receiver.

SAW - Surface Acoustic Wave - A type of steep-skirted filter used in the baseband or IF section of satellite reception and transmission equipment.

SCADA-Satellite Collection and Data Access. This refers to the use of Ultra Small Aperture Terminals installed in rural and remote areas to collect and transmit data such as for pipelines, small electric power generators, etc.

Scalar Feed - A type of horn antenna feed which uses a series of concentric rings to capture signals that have been reflected toward the focal point of a parabolic antenna.

SCORE - This was the first communications satellite launched by the United States. It was designed and launched on December 18, 1958 by the US. Signal Corps. It stored a message from President Eisenhower that was broadcast worldwide: "Peace on Earth good will to men." The acronym stood for Signal Communications for Orbiting Relay Equipment.

Scrambler - A device used to electronically alter a signal so that it can only be viewed or heard on a receiver equipped with a special decoder.

Secam - A color television system developed by the French and used in the Francophile countries and the USSR. Secam operates with 625 lines per picture frame and 50 cycles per second, but is incompatible in operation with the European PAL system or the US. NTSC system.

Sidelobe- Off-axis response of an antenna.

Signal to Noise Ratio (S/N) - The ratio of the signal power and noise power. A video S/N of 54 to 56 db is considered to be an excellent S/N, that is, of best broadcast quality. A video S/N of 48 to 52 Db is considered to be a good S/N at the headend for Cable TV.

Simplex Operation - Transmissions sent alternately in each direction in a telecommunications channel.

Single-Channel-Per-Carrier (SCPC) - A special high power audio service used to transmit a large number of signals over a single satellite transponder. Although far fewer channels are handles (a transponder normally has the capacity to carry 2000 voice conversations), each signal is transmitted with far more power, thereby allowing much smaller receive-only antennas to be used at the earth stations. Using SCPC techniques, the American news wire services are transmitting audio news feeds to radio stations nationwide equipped with antennas only three to four feet in diameter.

Single Sideband (SSB) - A form of amplitude modulation (AM) whereby one of the sidebands and the AM carrier are suppressed.

Skew - An adjustment that compensates for slight variance in angle between identical senses of polarity generated by two or more satellites.

Slant Range - The length of the path between a communications satellite and an associated earth station.

Slot - That longitudinal position in the geosynchronous orbit into which a communications satellite is "parked". Above the United States, communications satellites are typically positioned in slots which are based at two to three degree intervals.

SMATV (Satellite Master Antenna System) - The adding of an earth station to a MATV system to receive satellite programs.

Snow - A form of noise picked up by a television receiver caused by a weak signal. Snow is characterized by alternate dark and light dots appearing randomly on the picture tube. To eliminate snow, a more sensitive receive antenna must be used, or better amplification must be provided in the receiver (or both).

Solar Array - A network of solar cells which generate electricity from sunlight.

Solar Outage - Solar outages occur when an antenna is looking at a satellite, and the sun passes behind or near the satellite and within the field of view of the antenna. This field of view is usually wider than the beam width. Solar outages can be exactly predicted as to the timing for each site.

Sparklies - A form of satellite television "snow" caused by a weak signal. Unlike terrestrial VHF and UHF television snow which appears to have a softer texture, sparklies are sharper and more angular noise "blips". As with terrestrial reception, to eliminate sparklies, either the satellite antenna must be increased in size, or the low noise amplifier must be replaced with one which has a lower noise temperature.

Spectrum - The range of electromagnetic radio frequencies used in transmission of voice, data and television.

Spillover - Satellite signal that falls on locations outside the beam pattern's defined edge of coverage.

Spin Stabilization - A form of satellite stabilization and attitude control which is achieved through spinning the exterior of the spacecraft about its axis at a fixed rate.

Splitter - A passive device (one with no active electronic components) which distributes a television signal carried on a cable in two or more paths and sends it to a number of receivers simultaneously.

Spot Beam - A focused antenna pattern sent to a limited geographical area. Spot beams are used by domestic satellites to deliver certain transponder signals to geographically well defined areas such as Hawaii, Alaska and Puerto Rico.

Spread Spectrum - The transmission of a signal using a much wider bandwidth and power than would normally be required. Spread spectrum also involves the use of narrower signals that are frequency hopped through various parts of the transponder. Both techniques produce low levels of interference between the users. They also provide security in that the coded signals appear as though they were random noise to unauthorized earth stations. Both military and civil satellite applications have developed for spread spectrum transmissions.

Sputnik 1 - The first artificial satellite ever launched. This satellite was launched by the Soviet Union in October, 1957 for scientific measurements.

SSMA - Spread spectrum multiple access. Refers to a frequency multiple access or multiplexing technique. (See above).

SSPA - Solid state power amplifier. A VSLI solid state device that is gradually replacing Traveling Wave Tubes in satellite communications systems because they are lighter weight and are more reliable.

Starsys - A proposed low earth orbit messaging satellite system which is backed by French aerospace interests. It would compete with the OrbCom satellite system.

Stationkeeping - Minor orbital adjustments that are conducted to maintain the satellite's orbital assignment within the allocated "box" within the geostationary arc.

Subscription Television (STV) - A broadcasting television station transmitting pay television (usually first-run movies) in a scrambled mode.

Subcarrier - A second signal "piggybacked" onto a main signal to carry additional information. In satellite television transmission, the video picture is transmitted over the main carrier. The corresponding audio is sent via an FM subcarrier. Some satellite transponders carry as many as four special audio or data subcarriers whose signals may or may not be related to the main programming.

Subsatellite Point - The unique spot over the earth's equator as-signed to each geostationary satellite.

Superband - The frequency band from 216 to 600 MHz, used for fixed and mobile radios and additional television chan-nels on a cable system.

Superstation - A term originally used to describe Ted Turner's WTBS

Channel 17 UHF station in Atlanta, Georgia. With the addition of several other major independent television stations whose signals are also carried by a satellite in the US., the term superstation has come to mean any regional television station whose signal is picked up and retransmitted by satellite to cable companies nationwide.

Synchronization (Sync) - The process of orienting the transmitter and receiver circuits in the proper manner in order that they can be synchronized. Home television sets are synchronized by an incoming sync signal with the television cameras in the studios 60 times per second. The horizontal and vertical hold controls on the television set are used to set the receiver circuits to the approximate sync frequencies of incoming television picture and the sync pulses in the signal then fine-tune the circuits to the exact frequency and phase.

Syncom - The series of experimental satellites designed by NASA and Hughes Aircraft that demonstrated the technical feasibility of geosynchronous satellite operation. The first satellite in the series Syncom 1 was a launch failure. The second satellite, Syncom 2, launched in 1963 was fully successful in achieving and operating in geosynchronous orbit. It served as the prototype for the first operational geosynchronous satellite, Early Bird or INTELSAT 1, launched in April, 1965.

T

T1 - The transmission bit rate of 1.544 millions b/s per second. This is also equivalent to the ISDN Primary Rate Interface for the US. The so-called European T-l transmission rate is 2.048 million bits per second.

T3 Channel (DS-3) - In North America, a digital channel which communicates at 45.304 Mbps.

TDMA - Time division multiple access. Refers to a form of multiple access where a single carrier is time shared by many users. Signals from earth stations reaching the satellite consecutively are processed in time segments without overlapping.

Teleconference - An electronic multilocation, multiperson conference using audio, computer, slow-scan, or full-rate video systems.

Teledesic - The name of the US. proposed mega-LEO satellite system that would deploy 840 satellites for global telecommunications services. This system would deploy 40 satellites in each of 41 planes and each satellite would interconnect via intersatellite link with the 8 satellites that are most closely connected in the global orbital grid. This system would operate in the Ka Band (30/20 GHz) and would very intensively re-use this frequency to provide fixed, mobile and rural services at rates ranging from Kb/s to Gb/s. It is backed by Bill Gates of Microsoft and Craig McCaw of McCaw Cellular Inc.

Teletext - An "electronic newspaper of the air" often consisting of several hundred "pages" of 24 character by 20 line English text displays. Teletext can be used to cover several different applications. Most typically it is a one-way broadcast of preset information that is formatted as a "video-magazine" with information being accessed off of a table of contents. Teletext information can be sent simultaneously with a video signal over a satellite transponder using a data subcarrier. Several satellite programmers are now transmitting teletext signals, and the national television networks have begun to imple-ment their own teletext systems. A special teletext decoder is required to capture and display the teletext pages as they are transmitted. Three interactive teletext systems known view data or videotext systems now exist. These are the British Prestel, the French Antiope or Minitel and the Canadian Telidon systems. The interactive systems require the use of telephone lines to obtain the specially ordered information as opposed to the preset information available via teletext.

Telstar - The first satellite to demonstrate wide band broadcast relays. This experimental satellite transmitted a televi-sion signal upon its launch in 1962. It was designed and built by AT&T Bell Labs. The AT&T Corporation has maintained its trademark for the Telstar name and currently operates its domestic satellite system under the Telstar name.

Terrestrial TV - Ordinary "over the air" VHF (very high frequency) and UHF (ultrahigh frequency) television transmissions which are usually limited to an effective range of 100 miles or less. Terrestrial tv transmitters operate at frequencies between 54 megahertz and 890 megahertz, far lower than the I4/I2 and 6/4 billion hertz (gigahertz) microwave frequencies used by satellite transponders.

Three-Axis Stabilization - Type of spacecraft stabilization in which the body maintains a fixed attitude relative to the orbital track and the earth's surface. The reference axes are roll, pinch, and yaw, by nautical analogy.

Threshold Extension - A technique used by satellite television receivers to improve the signal-to noise ratio of the receiver by approximately 3 db (50%). When using small receive-only antennas, a especially equipped receiver with a threshold extension feature can make the difference between obtaining a decent picture or no picture at all.

Thruster - A small axial jet used during routine stationkeeping activities. These are often fueled bydrazine or bi-propellant. In time ion-engines will probably replace such thrusters.

TI - Terrestrial Interference - Interference to satellite reception caused by ground-based microwave transmitting stations.

Tongasat - A corporation established in the Kingdom of Tonga. This corporation has filed for a number of satellite locations in geosynchronous orbit and in multiple freqencies. These filings with the ITu have subsequently been leased to other satellite systems such as Unisat and Rimsat. This has been characterized as a "Flag of Convenience" approach of acquiring geosynchronous satellite locations.

Transfer Orbit - A highly elliptical orbit which is used as an intermediate stage for placing satellites into geostationary orbit.

Transmitter - An electronic device consisting of oscillator, modulator and other circuits which produce a radio or television electromagnetic wave signal for radiation into the atmosphere by an antenna.

Transponder - A combination receiver, frequency converter, and transmitter package, physically part of a communications satellite. Transponders have a typical output of five to ten watts, operate over a frequency band with a 36 to 72 megahertz bandwidth in the L, C, Ku, and sometimes Ka-Bands or in effect typically in the microwave spectrum, except for mobile satellite communications. Communications satellites typically have between 12 and 24 on-board transponders although the INTELSAT VI at the extreme end has 50.

Transponder Hopping - A single TDMA equipped earth station can extend its capacity by having access to several downlink beams by hopping from one transponder to another. In such a configuration the number of available transponders must be equivalent to the square of the number of beams that are interconnected or cross-strapped.

TSS - Telecommunications Standardization Sector. The world standards setting organization resulting from the combination of the CCITT (Consultative Committee on Telephony and Telegraphy) and the CCIR (Consultative Committee on International Radio).

Turnkey - Refers to a system that is supplied, installed and sometimes managed by one vendor or manufacturer.

TVRO - Television Receive Only terminals that use antenna reflectors and associated electronic equipment to receive and process TV and audio communications via satellite. Typically small home systems.

Tweeking - The process of adjusting an electronic receiver circuit to optimize its performance.

Two-way Capacity (Bi-Directional) - A Cable Television system with two-way capacity which can conduct signals to the headend as well as away from it. Two-way or bi-directional systems now carry data; they may eventually carry full audio and video television signals in either direction.

TWT - Travelling-wave tube.

TWTA - Travelling-wave tube amplifier.

U

Ultrahigh Frequency (UHF) - Officially the band of frequencies ranging from 300 to 3000 MHz. In television use, refers to the set of frequencies starting at 470 MHz. The UHF channels are designated as 14 through 70.

Uplink -The earth-to-space telecommunications pathway.

USAT- Ultra Small Aperture Terminal. This refers to very small terminals for SCADA, DBS and other satellite applications where the terminal can be very small as in under 50 cms. The new Hughes Spaceway satellite system would employ USATs for fixed satellite services. Similar size terminals could be used on the Japanese NSTAR DBS system and the Teledesic Mega-LEO satellite system.

V

Value-Added Carrier - Carrier that orders services from common carriers, then adds special features before retailing the use of these circuits. Value-added carriers supply computer-oriented services. Service provided is also known as a value-added network (VAN).

Van Allen radiation belts - These are two high level radiation belts discovered by an Explorer Satellite designed by Dr. Van Allen of Cal Tech. These belts which are highly destructive to communications satellites consists of two belts of highly charged particles and high energy neutrons.

VBI- Vertical blanking interval.

Vertical Interval Test Signal - A method whereby broadcasters add test signals to the blanked portion of the vertical interval. Normally placed on lines 17 through 21 in both field one and two.

Very High Frequencies (VHF) - The range of frequencies extending from 30 to 300 MHz; also television channels 2 through 13.

VSAT - Very small aperture terminal. Refers to small earth stations, usually in the l.2 to 2.4 meter range. Small aperture terminals under 0.5 meters are sometimes referred to Ultra Small Aperture Terminals (USAT's)

W

WARC - World Administrative Radio Conference.

Waveguide - A metallic microwave conductor, typically rectangular in shape, used to carry microwave signals into and out of microwave antennas.

X

X-Band - The frequency band in the 7-8 GHz region which is used for military satellite communications

X.25 - A set of packet switching standards published by the CCITT.

X.400 - A set of CCITT standards for global messaging.

Z

Zulu Time - This is the same a Greenwich Meridian Time (GMT). This is the time standard used in global satellite systems such as INTELSAT and INMARSAT in order to achieve global synchronization.

<u>SELECTED BIBLIOGRAPHY</u>

<u>Books</u>

Alper, Joel and Pelton, Joseph N. (editors). The INTELSAT Global Satellite Network (New York: AIAA Progress Series, 1986).

1993-94 Annual Review of Communications (Chicago, Illinois: International Engineering Consortium, 1994)

Branscomb, Anne (ed.), Toward a Law of Global Communications Networks. (New York: Longman, 1986).

Bruce, Robert, Cunard, Jeffrey and Director, Mark., The Telecommunications Mosaic (London, Butterworth Scientific, 1988).

Didsbury, Howard (ed.), The Future: Opportunity, Not Destiny (Bethesda, Maryland: World Future Society, 1986)

Feher Kamilo, Digital Communicatiosn: Satellite/Earth Station Engineering. (Englewood Cliffs, New Jersey: Prentice Hall, 1983)

Hills, Jill, Deregulating Telecommunications. (London: Frances Pinter Publishing, 1986).

Howkins, John and Pelton, Joseph N., Satellites International. (London: Macmillan Press. Ltd., 1986)

Hudson, Heather, Communications Satellites: Their Development and Impact. (New York: Free Press, 1990)

Keen, Peter G. W. and Cummins, J. Michael, Networks in Action: Business Choices and Telecommuncations Decisions. (Belmont, California: Wadsworth Publishing, 1994)

Meyers, Richard ed., The Encyclopedia of Telecommunications. (San Diego, California: Academic Press, 1989).

Meagher, Christine, Satellite Regulatory Compendium. (Potomac, MD.: Phillips Publishing, 1993)

Mirobio, M. M. and Morgenstern, Barbara The New Communications Technologies. (Boston, Ma.: Focal Press, 1990)

Miya, K. (editor), Satellite Communications Technology. (Tokyo, Japan: Institute of Electronic and Communication Engineers of Japan, 1985)

National Research Council, Computer Science and Teleocmmunications Board, Realizing the Informaiton Future: The INTERNET and Beyond. (Washington, D.C., National Research Council, 1994)

Newton, H., Newton's Telecommunicaitons Dictionary, (New York: Telecommunications Library, 1993)

Payne, Silvano (ed.), International Satellite Directory. (Sonoma, CA.: Design Publishers, 1995)

Pelton, Joseph N., Future View: Communications, Technology and Society, (Boulder, Colorado: Baylin Publications, 1992)

Pritchard, Wilbur, Suyderhoud, Henri, and Nelson, Robert. Satellite Communication Systems Engineering, (Englewood Cliffs, New Jersey: Prentice Hall, 1993)

Schnaars, Steven, Megamistakes (New York: MacMillian, 1986).

U.S. Congress, Office of Technology Assessment, Critical Connections: Communications for the Future. (Washington, D.C., OTA, 1990).

Williams, Fred, The New Telecommunications: Infrastructure For the Information Age (New York: The Free Press, 1991)

Williamson, Mark, The Communications Satellite. (Bristol, United Kingdom: Adam Hilger, 1990)

Articles

George Abe "The Global Network" Network Computing November 15, 1993 pp 48-53.

Jeannine Aversa, "FCCs Okays Satellites Providing Two Way Data Services" Daily Camera, October 1994.

Walter S. Baer, "New Communications Technologies and Services," in Paula R. Newberg, ed., New Directions in Telecommunications Policy (Durham, N.C.: Duke University Press, 1989) 139-169.

"Band Sharing, 6; Band Splitting 1: Big LEO Report Sent to FCC" SIGNALS. No. 6 Spring /Summer 1993 pp. 1.

William T. Brandon, "Market Elasticity of Satellite Communications Terminals" Journal of Space Communications, Vol 10, No. 4. pp. 279-284.

Steve Brown. "How the Europeans Respond to Mobile Communications", Communications News. May 1994 pp. 28-30.

Jane Bryant, "Mobile Satellite Services: an Update" Via Satellite. October, 1993 pp. 48-52.

Simon Bull, "The Status of the Interactive VSAT Market" Via Satellite. December 1994 pp. 34-38.

Chris Bulloch, "A Cautiously Optimistic Look at European VSATs" Via Satellite. December 1994 pp.40-43.

John Carey and Mitchell Moss, "The Diffusion of New Telecommunications Technologies", Telecommunications Policy. June 1985 pp. 145-158.

Mary Lu Carnevale, "Broadcasters Gain Support for Measure to Open Spectrum for New Services" Wall Street Journal, March 1, 1994 pp. 6.

R.L. Carraway, J.M. Cummins, and J. R. Freeland, "The Relative Efficiency of Satellites and Fiber-Optic Cables in Multipoint Networks", Journal of Space Communications Amsterdam, Netherlands: IOS Press, Vol. 6, No. 4, January 1989 pp. 277-289.

"The Communicopia Study: C-5 Convergence", New York: Goldman-Sachs, 1992.

"Comsat Appeals for Privatization", Satellite Communications. (April,1994) pp. 10-11.

Allen Davis "Cable Overbuild: Alternative Video Access Opportunity", 1993-94 Annual Review of Communications (Chicago, Illinois: International Engineering Consortium, 1994) pp. 115-118.

Deloitte and Touche, "1993 Wireless Communications Industry Survey" (Atlanta, Georgia: Deloitte and Touche, 1993).

"Electronic Privacy Bill Passes House Committee" Telephony. (May 26, 1986)

Richard V. Ducey and Mark R. Fratrik, "Broadcasting Response to New Technologies", Journal of Media Economics, Fall, 1989 p. 80.

Rob Frieden, "Wireline vs. Wireless: Can Network Parity Be Reached?" Satellite Commuications.
July, 1994. pp. 20-23.

Peter Glaser, "The Practical Uses of High Altitude Long Endurance Platforms" International Astronautical Federation, Graz, Austria, October, 1993.

"Global 2000 Report on Telecommunications" (Washington, D.C.: National Telecommunications and Information Administration, 1990)

Green Paper on Satellite Communications, European Commission. Brussels, Belgium: The European Commission, 1992)

Carolyn Horwitz, "The Rise of Global VSAT Networks, Satellite Communications (Denver, Colorado: Argus Publishing, April 1994). pp. 31-34.

Takeshi Iida, "Designing Low Cost Satellite Communications Systems for Remote Tele-Science", Communications Research Laboratory. 1993.

The INTELSAT Agreement and Operating Agreement,TIAS Series (Washington, D.C., U.S. State Department, 1992)

ITU International Table of Frequency Allocations as contained in Part 2 of the FCC Rules and Regulations (Wasington, D.C.: Government Printing Office, 1994).

"Ku-Band Sharing" Satellite Communications (April 1994) pp.12.

Robert Lang and Jon Sauer, "Scalable Dense Wave Division Multiplex Photonics for an All Optical Network", White Paper, Spectra Diode Laboratory, San Jose, California, 1994)

Gerard Maral, "Comparative Analysis of LEO, MEO, and GEO Systems", Lecture at the International Space University, Barcelona, Spain, August 1994.

M. Lynne Markus, "Toward a Critical Mass Theory of Interactive Media: Universal Access, Interdependence and Diffusion", Communications Research, October, 1987. pp. 491-210.

Peter Marshall, "Global Television by Satellite" The Journal of Space Communications and Broadcasting. Vol 6, No. 4 , January 1989.

C. Mason, Will the U.S. Remain Competitive in the Wireless Future" Telephony (July 12, 1993)

Craig O. McCaw "Cellular Communications, 1993-94 Annual Review of Communications (Chicago, Illinois: International Engineering Consortium, 1994) pp. 43-44.

Patrick McDougal and Victor Barendse, "INMARSAT and Personal Mobile Satellite Services" Journal of Space Communications. Vol. 11, No. 2 pp. 2-11.

"Motorola's Pioneer Perference Applicaton for Iridium Loses" FCC Week (September 1992).

Oberst, Gerald E. Jr. "FCC Reorganization Puts Satellites in the International Bureau" Via Satellite. December 1994. pp.14-15

L. M. Paschall, "Security aspects of Satelite and Cable Systems" Journal of Space Communications Amsterdam, Netherlands: IOS Press, Vol. 6, No. 4, January 1989. pp. 269-276.

Joseph N. Pelton, "Five Ways Nicholas Negroponte is Wrong About the Future of Telecommunications" Telecommunications (Vol. 11, No. 4, April 1993.

Joseph N. Pelton, "The Globalization of Universal Telecommunications Services" in Universal Telephone Service: Ready for the 21st Century? (Wye, Maryland, Institute for Information Studies and the Aspen Institute Annual Review, November 1991, pp.141-151.

Joseph N. Pelton, "How INTELSAT Was Privatized While No One Was Looking?" Via Satellite (February, 1989).

Joseph N. Pelton, "Toward a New National Vision for the Information Highway" Telecommunications (Vol. 11, No. 9) September 1993.

Joseph N. Pelton, "Will the Small Satellite Market Be Large" Via Satellite (April, 1993).

Joseph N. Pelton "Overview of Satellite Communications: Global Trends in Deregulation and Competition", International Astronautical Federation, 45th Congress, Jerusalem, Israel, October, 1994.

"Public's Privacy Concerns Still Rising", Privacy and American Business. (Hackensack, N.J.: Center for Social and Legal Research, September, 1993)

"Reformed ITU Filing Procedures—Brokers for orbital space-boon or bane?" Via Satellite , May 1994.

Richard Jay Solomon, "Shifting the Locus of Control" Annual Review of Communications and Society (Queenstown, MD: Institute for Information Studies, 1989).

Special Edition on Mobile and Small Satellites: Journal of Space Communications (Vol. 10, No. 2), April 1993.

Lynda B. Starr, "NASA ACTS: Exploring the Next Generation of Satellite Users", Wireless. Volume 3 No. 4 July/Augst 1994, pp. 42-43.

James Stuart, "Microsatellites Comes of Age" Journal of Space Communications. Vol. 10 No. 1. pp.3-8.

Leslie Taylor, "PCS Frequency Auction", Signals, November 1993.

"The Information Wave: Digital Compression is Expanding the Definition of Modern Telecommunications" Uplink (Los Angeles, California: Hughes Aircraft, Spring, 1994) pp. 4-7.

Roy A. Wainwright, "Quality as a Competitive Edge", 1993-94 Annual Review of Communications (Chicago, Illinois: International Engineering Consortium, 1994) pp. 939-943.

"World's Top 10 Markets for Telecommunications Investment Forecasted" Global Telecom Report (Vol. 4, No. 7) April 4, 1994. pp. 1-2.

Dinah Zeiger, "Four Team Up For Wireless Service" Denver Post. October 21. 1994. pp.1C and 10C.

Reports and Documents

Center for Telecommunications Management, The Telcom Outlook Report. Chicago, Illinois: International Engineering Consortium, 1994)

Federal Communications Commission, "Regulatory Treatment of Mobile Services: Report and Order on Personal Communications Services" (Washington, D.C.: Adopted June, 1994)

Federal Communications Commission, "Regulatory Treatment of Mobile Services: Third Report and Order Related to Specialized Mobile Radio Services" (Washington, D.C.: Adopted August 9, 1994 and released September 23, 1994).

International Telecommunication Union. "ITU-T Recommendation F.850— Principles of Univrsal Personal Communications. (March 1993)

NASA / NSF, Panel Report on Satellite Communications Systems andTechnology, Volumes 1 and 2. (Baltimore, Maryland: International Technology Research Institute, July, 1993)

═══ INDEX ═══

A

B

● ● ● *books*
> ● ● ● *databases*
>> ● ● ● *training videos*
>>> ● ● ● *news services*

> **Complete List of Design Publishers'**
> **Satellite Communications Products**

● ● ● *books*

INTERNATIONAL SATELLITE DIRECTORY
$260.00

> The most complete reference source available in the
> world on communications satellites. Includes all manu-
> facturers of ground and space equipment, service pro-
> viders & users of satellite services, international regu-
> lators and satellite operators. Footprint maps of all
> satellites including detailed satellite technical specifica-
> tions. Updated annually with 14 separate chapters &
> over 25,000 entries, this best selling 1350 plus page
> book is a must for all satellite professionals.
>
> Also available in CD-ROM format - see later
>
> *Postage & Handling: USA - $8.50 All Other Air - $45.00*
> *Surface - $15.00*

WHAT'S ON SATELLITE - (Annual Subscription)
$125.00

> Three times a year receive this report on video, voice
> and data activity on all the world's satellites. The
> polarity, bandwidth, frequencies, video & radio pro-
> grams, SCPC, EIRP levels & much more, is listed in
> detail for each satellite. This information allows you to
> keep abreast of new programming, satellite launches,
> new services and satellite relocations.
>
> *Postage & Handling: USA - $8.50 All Other Air - $35.00*

NEWNES GUIDE TO SATELLITE TV by D.J. Stephenson
$ 40.00

Newnes Guide to Satellite TV is a practical guide, without excessive theory or mathematics, to the installation and servicing of satellite TV receiving equipment for those professionally employed in the aerial rigging and TV trades. It continues to meet that practical need between theoretical text book and simple installation guide. New material includes a rewritten link budget chapter, which gives a detailed calculation method allowing for operational losses and digital extensions.

Postage & Handling: USA - $3.50 All Other Air - $10.00

WORLD SATELLITE TV AND SCRAMBLING METHODS - by Dr. Frank Baylin
$ 45.00

The design, operation and repair of satellite antennas, feeds, LNB's & receivers/modulators are examined in detail. An in-depth study of scrambling methods from VideoCypher, Oak Orion, FilmNet, Sky Channel, EuroCypher, D2 MAC, and other systems. Circuit and block diagrams of all components are presented and clearly explained throughout the handbook.

Postage & Handling: USA - $4.50 All Other Air - $12.00

HOME SATELLITE TV INSTALLATION & TROUBLESHOOTING MANUAL - 3rd Edition Revised - by Dr. Frank Baylin
$ 40.00

An invaluable sourcebook for owners of home satellite TV systems and professional installers alike, it thoroughly explores the fundamentals of satellite communications and earth station operation, explains how to select and evaluate satellite television components, details installation procedures including multiple-receiver and multiple-television hookups and large antennas, and sets out a complete strategy for troubleshooting any satellite TV system.

Postage & Handling: USA - $4.50 All Other Air - $15.00

MINIATURE SATELLITE DISHES - The new Digital Television
by Dr. Frank Baylin
$ 25.00

150 channels of satellite television from 18' dishes! Learn all about the players, the technology & the programming. Features photographs and information supplied directly from DirectTV, Primestar, Hubbard Communications/USSB & Fortuna Communications.

Postage USA - $3.50 All Other Air - $9.00

KU-BAND SATELLITE TV - by Dr. Frank Baylin
$ 35.00

A clear presentation of small dishes, DBS satellite TV, flat plate antennas, actuators, LNB and satellite receiver electronic design, worldwide scrambling technologies, link analysis, fixed antenna installation, interfacing receivers and decoders and aligning polar mounts.

Postage & Handling: USA - $4.50 All Other Air - $12.00

● ● ● *databases*

● ● *CD-ROM*

INTERNATIONAL SATELLITE DATABASE-

$925.00

Available for IBM computers & compatibles running on Windows this is the complete database on satellites and companies in the industry. Contains all the information you need to operate successfully in the satellite industry. Price includes a copy of the International Satellite Directory (Value $260).

Postage & Handling without Directory: Free

Postage & Handling with Directory: USA - $9.50 All Other Air - $45.00, Surface - $15.00

•• *INTERNET & ONLINE*

SATNEWS.COM

From - $295.00

The most complete database available through the INTERNET or online direct dial. You can lookup the following databases: <u>What's On Satellite:</u> This report lists what video, voice and data traffic is on each of over 90 communications satellites. Details include the name of the program, transponder number & frequency, audio subcarriers etc. <u>Satellites:</u> Over 30 different fields of information on over 400 satellites. Includes details on owners & operators, geosynchronous position, type of satellite, launch details, manufacture, frequencies with EIRP, SFD and G/T information on most of the world's operating satellites. <u>Companies & Organizations:</u> Search for information on close to 10,000 companies and organizations involved in satellite communications.

••• *training videos and courses*

SATELLITE TRAINING SERIES - (Series of 12 VHS Video Tapes)

$75.00 to $595.00

Twelve videos which include the following subjects:- 1. Beginners Briefing on Satellite Communications; 2. History and Organizations; 3. Satellite and Antenna Technology Basics; 4. Satellite Fundamentals; 5. The World Satellite Marketplace; 6. Financial Aspects; 7. Earth Stations and Terminals; 8. Maintenance of Commercial Earth Stations; 9. Description of Various Satellite Services; 10. Basic Telecommunication Principles; 11. VSAT Systems and Management; 12. Future Trends. - For a complete color brochure please contact Design Publishers

IEEE SATELLITE COMMUNICATIONS COURSE
$325.00

A solid self study course offered by the "Institute of Electrical & Electronics Engineers" for the advanced technical person. You'll learn to: achieve the required performance, design, and analysis of digital satellite communications links, understand the spacecraft that make satellite communication possible and the launch vehicles that put them in orbit, understand and use basic communication theory, link equations, modulation techniques, digital communication fundamentals, evaluate what is needed to make a satellite communication system, put together a satellite communications link, understand the earth station that sends and receives signals, large and small antennas. An IEEE Certificate of Achievement will be awarded on successful completion of the final exam. *Postage & Handling: USA - $5.50 All Other Air - $25.00*

● ● ● *news services*

SATNEWS.COM (Price depends on usage)

From $295.00

The most complete satellite news service available through the INTERNET or online direct dial. You can access the following information on a daily basis: <u>News in Detail</u>: A comprehensive, in-depth and timely source for news, business news and financial reports on the entire satellite industry. Provides up-to-the-minute information on business and economic developments from major corporations and governments. Covers new contract awards, product announcements, earnings reports, personnel shifts, and more. In addition to news stories, regular offerings include analysis, reviews and ratings and full-text press releases. Updated daily by our editorial staff. <u>Keyword Search</u> Provides all the news since the start of the service - thousands of articles that you can search for exactly what you want! <u>E-Mail</u>: Provides a place for the entire industry to exchange E-Mail. SATNEWS•COM is an official Internet address so anyone can send mail to you - and likewise you can send them mail through the Internet. <u>Calendar of Events</u>: Lists events, conferences and exhibitions scheduled by key private sector and government bodies in the satellite industry. Provides crucial advance notice of forthcoming events and up-to-date details of time and location. <u>Legal & Regulatory Reports</u>: Provides details on Congressional hearings, conferences, and press briefings of key FCC and other governmental officials and bodies in Washington when discussing communications issues. Provides coverage of the International Telecommunications Union, IFRB and other regulatory bodies and newsmakers. <u>Financial Reports</u>: Provides key figures on the financial aspects of both public and some private companies involved in satellite communications. Stock quotes, securities offerings, M & A, dividends, earnings reports, new contracts etc.

ORDER FORM

SEND TO:-   **DESIGN PUBLISHERS**
800 Siesta Way • Sonoma
CA 95476 • USA

FOR FASTER SERVICE CALL OR FAX
Tel: (707) 939-9306 Fax: (707) 939-9235

Quantity	Product Description	Price	Total
	Please add on Postage & Handling		
	TOTAL		

Name:_____

Company:_____

Address:_____

City:_____ State:_____ Zip: _____

Country:_____

Telephone: _____   Fax: _____

_____ **Payment Enclosed:** Make check payable in US$ drawn on a
US bank to "Design Publishers"

_____ **Bill Company**. PO No. _____ Product provided
on receipt of payment.

_____ **Wire Transfer** to: Bank of America, Sonoma Account
No. 1043-01485. Please add $10.00 transfer fee and ensure
your bank includes your Company name for reference.

_____ **Charge** the following card:
VISA _____ Master Card _____ AMEX _____

Card Number:_____

Expiration Date:_____

Signature:_____